Diversity and Function in Mangrove Ecosystems

Developments in Hydrobiology 145

Series editor
H. J. Dumont

Diversity and Function in Mangrove Ecosystems

Proceedings of Mangrove Symposia held in Toulouse, France, 9–10 July 1997 and 8–10 July 1998

Edited by

R.S. Dodd

Reprinted from Hydrobiologia, volume 413 (1999)

Springer Science+Business Media, B.V.

Library of Congress Cataloging-in-Publication Data

A C.I.P. Catalogue record for this book is available from the Library of Congress.

ISBN 978-0-7923-6158-9 ISBN 978-94-011-4078-2 (eBook)
DOI 10.1007/978-94-011-4078-2

Printed on acid-free paper

Contents

Hydrobiologia **413:** vii, 1999.
R.S. Dodd (ed.), Diversity and Function in Mangrove Ecosystems.

Preface

The tropical coastal mangrove forest, once considered to be an inhospitable insect-infested swamp, the pariah among ecosystems, but paradise for the developer, has taken a new image in recent years. Growing awareness of its role in the intricate web of marine life, its value as a natural resource for local populations and the protection it affords to easily eroded coastlines, has contributed to a surge of interest, both scientific and lay. Nevertheless, our knowledge of the evolution and functioning of these ecosystems remains superficial and substantial gaps in our understanding can be identified. An inadequate comprehension of the interrelationships between ecosystem functioning and the diversity and evolution of it's component parts has surely been the cause of disasters in rehabilitation of degraded mangrove ecosystems, or in afforestation of new lands. It is recognized that mangrove forests are being degraded and lost at an alarming rate, leading to local social and economic distress and to calls for conservation efforts by the international community. These concerns emphasize the need for future focused research directions.

Several earlier publications have arisen from symposia dedicated to mangrove research. However, the pace of advances in research techniques in recent years has opened up opportunities to revisit earlier questions and to develop new areas of inquiry. Notable here are the advances in molecular biology, that offer opportunities for a more profound knowledge of biosystematics of mangrove taxa, of species genetic architecture, of physiological processes and of biotechnological potential. Advances in imaging technology have revolutionized methods in vegetation cartography and temporal assessment of vegetation dynamics, allowing remote monitoring of the effects of local and global environmental change. Field ecology has been brought increasingly into the laboratory, where controlled cause and effect relationships can be compared with *in vivo* conditions. A revolution in thinking about biodiversity has led to debate on conservation and sustained ecosystem management, opening up the biological dimension to include the social.

Because of these advances, it was considered timely to bring together specialists in mangrove research to review present and future research. Two symposia were held at Toulouse, France. In July 1997 about 50 scientists from all continents with important mangrove resources participated in a symposium entitled “Mangrove Ecosystems: Biodiversity, Functioning, Restoration and Management”. This was followed in July 1998 by a second meeting entitled “Recent Advances and Future Trends in Mangrove Research: Biodiversity, Genetics, Evolutionary Biology and Restoration”. This special issue of the journal Hydrobiologia includes a selection of the papers that were presented at the two meetings. It should be considered an important contribution, though necessarily incomplete, covering various aspects of the mangrove ecosystem appropriate for the scientific community as well as being a useful source for wetland management.

At the end of the second meeting the participants convened as an *ad hoc* working group to identify priority areas for future research. Professor Colin Field, Dr Daniel Imbert and Professor Peter Saenger presented overviews of gaps in knowledge of mangrove ecosystems and recommended future research efforts. Professor Samuel Snedaker took responsibility for leading the group discussion that resulted in the statement “The Future of Mangrove Research”.

It is my pleasure as organizer of the symposia to thank all participants. My gratitude is extended to Director F. Blasco for scientific collaboration and for organizing financial support from the following sponsors: Centre National de la Recherche Scientifique (CNRS), Université Paul Sabatier, Toulouse, Région Midi-Pyrénées. My thanks also go to Kluwer Academic Publishers and Editor-in-Chief, Professor H. Dumont, University of Ghent, Belgium for agreeing to publish the Proceedings as a special volume in Hydrobiologia.

PROFESSOR RICHARD S. DODD
Guest Editor

Hydrobiologia **413**: ix–x, 1999.
R.S. Dodd (ed.), Diversity and Function in Mangrove Ecosystems.

The future of mangrove research

The Working Group represented ongoing research activities in Africa, Asia, Australia, the Middle East, and the Americas. During the course of the three-day symposium, prepared presentations were made on mangrove biodiversity, mangrove genetics, the evolutionary biology of mangroves, and the rehabilitation and restoration of degraded mangrove areas. Subsequent discussions on the various topics revealed a notable lack of a global consensus or unifying focus on the future of mangrove research. Whereas the lack of a global consensus might not limit new advances in mangrove research, the Group strongly felt that a common focus would promote interdisciplinary initiatives, accelerate research in promising areas, and lay a basis for enhanced global collaboration. Accordingly, the assembled Group convened as an *ad hoc* working Group to define future mangrove research priorities as well as operational research-support needs that are required to promote and sustain research in the priority areas. The resulting consensus is summarized below:

Molecular and evolutionary biology of mangrove systems

In terms of basic mangrove research, the Group recognized that work on the molecular biology of these tropical vascular halophytes lags far behind similar work being done on economic crop species and certain plant models, such as Arabidopsis. The principal research needs elaborated by the Group emphasized mangrove genetics, e.g., evolutionary phylogenetic relationships, genomic imprinting, transfer RNA mechanisms, germ plasm and transgenic research. Equally far behind researchers working in non- mangrove forest habitats is research on the evolutionary biology of mangrove systems. Although a large field in itself, the Group placed specific research priority on plant–plant and plant–animal co-evolution, e.g. the co-evolution of epiphytes such as orchids and bromeliads and their mangrove hosts with emphasis on inter-species physiological relationships, and the role of mangrove animal species in the evolution of mangrove systems.

Mangrove system dynamics

Superficially, the gross physiognomy of mangrove forests appears no different than other low-elevation tropical forests in similar climates. Mangroves, however, are adapted to very different habitat conditions, including but not limited to sediment anoxia, salinity, a marine-tide driven hydrology, and severe cyclonic events that help to create a highly dynamic geomorphic setting. Although mangroves appear superbly adapted, each of these factors also act as stressors that modify mangrove forest dynamics. To achieve a better understanding of the dynamics of this unique coastal forest system, the Group agreed that future research should focus on both short-term processes (e.g., colonization strategies, mechanisms of adaptation to stress, habitat-controlled inter-species interactions) and long-term processes (e.g., the structuring role of catastrophic disturbances, and mangrove community responses to accelerating changes in sea level, ultraviolet radiation, and the composition of atmospheric gases). Because mangrove forest dynamics operate on differing temporal and spatial scales, advanced techniques in remote sensing were considered to be the principal tool that integrates diverse research interests.

The human dimension

Throughout recorded history, mangrove systems have provided a subsistence base for people living in the tropical coastal zone. Over recent decades, however, this forest system has been increasingly exploited for commercial gain (e.g., intensive harvesting of wood products, large-scale conversion to agriculture and human settlement, and the controversial conversion or mangrove areas to problematic mariculture). Consequently, large areas worldwide are highly degraded and stand in need of rehabilitation and restoration. In this regard, there is a demonstrable paucity of social science research in mangrove areas particularly in the context of how local people are affected by the loss of mangrove habitats and resources. Since the extent and rate of loss are largely unknown, the Group gave high priority to enhanced global and regional monitoring using GIS mapping techniques. With regard to degraded areas, the Group identified several key research questions that need to be aggressively addressed. These are: (1) What factors lead to success versus failure in mangrove rehabilitation and restoration projects? (2) Can the introduction of "new" mangrove species facilitate the restoration of barren coastal habitats? (3) How can biodiversity be optimized relative to efforts to enhance forest productivity? (4) Since resource economists have variously valued mangrove habitats in the range of US$ 25,000–75,000 per hectare in terms of the natural services they provide, how can rehabilitation and restoration efforts be modified to maximize the economic benefit to local peoples?

Operational research-support

Finally, the Group recognized that these future trends in mangrove research will require enhanced global communication and cooperation among mangrove researchers. To achieve this, the Group recommended the establishment of an international internet web site that will provide an entré into the published and grey literature, advice on preferred methodologies in mangrove research and valuation methods, and opportunities for inter-personal networking among mangrove researchers. The inclusion of an all- encompassing internet mangrove data base was largely dismissed because of the generic absence of input-data quality control and the validation of the research that generated the input data. However, a modified database that focused exclusively on specific topics (e.g., mangrove area maps, species descriptions along with diagnostic color photographs) would have broad appeal to the general public as well as mangrove researchers.

PROFESSOR SAMUEL C. SNEDAKER,
Group Moderator

Hydrobiologia **413:** 1–9, 1999.
R.S. Dodd (ed.), Diversity and Function in Mangrove Ecosystems.

A case of natural selection in Atlantic-East-Pacific *Rhizophora*

Zara Afzal-Rafii[1], Richard S. Dodd[2,*] & Marie-Thérèse Fauvel[3]
[1]*Institut Méditerranéen d'Ecologie et Paléoécologie, UA 1152, Université d'Aix-Marseille III, 13397 Marseille, France*
[2]*Laboratoire d'Ecologie Terrestre, UMR 5552, Université Paul Sabatier, BP 4403, 31405 Toulouse, Cedex, France*
[3]*Laboratoire d'Ecologie Terrestre, UMR 5552, 13 Avenue du Colonel Roche, BP 4403, 31045 Toulouse, Cedex, France*
(**Author for correspondence*)

Key words: ecological genetics, adaptation, *Rhizophora*, mangroves, cuticular waxes

Abstract

We have examined possible adaptation in cuticular alkane composition in the halophytic mangrove genus *Rhizophora*. Relative composition of the dominant alkanes varied: 1. among the three sympatric species from the Atlantic-East-Pacific region, 2. with geographic region within species and, 3. among populations within geographic region. For *R. mangle*, longer chain alkanes were more important in the semi-arid regions of north-west Africa and the Pacific coast of north-central Mexico. Mantel tests showed that inter-population taxonomic distances for the major alkanes were correlated with taxonomic distances for annual rainfall and mean maximum temperature, but not with weighted geographic distance. Since alkane carbon chain length should affect the biophysical properties of waxes, with longer chain lengths increasing crystallinity and impermeability, our data provides support for the hypothesis that observed differentiation is due to natural selection, rather than stochastic processes in this species. The same pattern was not observed for *R. racemosa* or *R. harrisonii*. Since these two species occupy less saline conditions and are more restricted in their latitudinal range, selection pressure may be less important than other evolutionary forces such as genetic drift. There was some evidence that alkane composition was more closely correlated with mean minimum temperatures in *R. racemosa*, that might set the latitudinal limits in this species.

Introduction

The relative roles of natural selection and stochastic processes in differentiation and speciation is of major interest to evolutionary biologists. Although phenotypic differences among species and populations can be observed for a wide range of characters, demonstration that such differences are adaptive can be elusive. The boundary layer between organs and the atmosphere, cuticular wax, which is made up of a complex mix of compounds, is thought to be important in controlling permeability and is a likely candidate for the forces of natural selection. Although cuticular waxes comprise a mixture of compounds, it is the long-chain hydrocarbons that are thought to be crucial in reducing permeability, due to their tight crystalline structure (Lockey, 1988; Reiderer & Schneider, 1990). Within the hydrocarbon fraction, straight chain *n*-alkanes are the most common constituents, with carbon chain length being related to biophysical properties of the wax mixture such as the transition melting temperature (Gibbs, 1995). Recently, Dodd et al. (1998) have shown a correlation between the relative proportions of alkanes of different carbon chain length and climatic conditions in natural populations of a South American conifer, with relative proportions of long chain alkanes increasing under more arid conditions. Preliminary results also indicate that cuticular wax composition may vary with aridity in the very widespread mangrove *A. marina* (Dodd et al., 1999). If cuticular wax composition is a genetically adaptive trait, it should be correlated with environmental variation, particularly in taxa occupying a wide geographic and climatic range and varying conditions of water stress, such as some mangrove taxa, including the genus *Rhizophora*.

Rhizophora is a pan-tropical genus with three species native to the Atlantic-East-Pacific region: *R.*

mangle L., *R. racemosa* G.F.W. Meyer and *R. harrisonii* Leechman. The three species are found on the West African and American Atlantic coasts (Keay, 1953); *R. mangle* and *R. harrisonii* are also known from the Pacific coast of America, with *R. racemosa* possibly being present (Prance et al., 1975; Gentry, 1982; Jiménez, 1987). Ecological differentiation among the three species has been described by several authors (Savory, 1953; Jonker, 1959; Breteler, 1969). *R. mangle* is considered to be the most salt tolerant of the three and is the only one to extend beyond the tropical belt. *R. harrisonii* is thought to be a hybrid between the two other Atlantic species (Breteler, 1969, 1977) and is commonly reported to occupy intermediate positions in the ecological zonation (Savory, 1953; Jonker, 1959). In earlier work, we showed significant variation in cuticular wax composition between West African and French Guianan populations of *Rhizophora* (A. Rafii et al., 1996) suggesting possible differentiation between New and Old World sources. However, lack of a broad range of sampling precluded testing for possible ecological adaptation in cuticular waxes. In this study, we have sampled populations of the three species from the humid tropics to semi-arid coastlines. In view of the greater ecological range of *R. mangle*, we have sampled this species most intensively to ask the questions: 1. Is there significant variation in the relative proportions of different alkane chain lengths among populations? 2. If so, is this variation correlated with variations in environmental conditions and, 3. Do environmentally correlated variations in alkane chain lengths correspond with the known biophysical properties of alkane mixtures? Habitat differentiation between *R. mangle* on the one hand and *R. racemosa* and *R. harrisonii* on the other, provides an opportunity to test for possible differential fitness of alkane profiles in closely related taxa.

Materials and methods

The three Atlantic east Pacific mangroves *Rhizophora mangle*, *Rhizophora racemosa* and *Rhizophora harrisonii* were sampled in localities from the west coast of Africa, the Atlantic coast of South America and North America, the Caribbean islands, and the Pacific coast of Mexico. Sampled populations are listed in Table 1. Foliage was collected from approximately 10 trees per population and either dried or kept in cool conditions before storage in a freezer.

Cuticular waxes were extracted by immersing 2–3 leaves in *n*-hexane for about 1.5 min. The extract was evaporated to dryness to remove water and redissolved in hexane before elution through a silica gel column (70–230 mesh). The hydrocarbon fraction that eluted from the column was injected into a Hewlett Packard® 5890 gas chromatograph equipped with a flame ionization detector onto an HT-5 capillary column (25 m long, 0.25 mm internal diameter). Oven temperature began at 160 °C, rising without delay by 10 °C min^{-1} to 240 °C, and then by 4 °C min^{-1} to 320 °C, with a final hold time of 5 min. Compounds were identified by comparison of retention times with commercial hydrocarbon standards and by GC-MS. Relative quantities of each hydrocarbon were calculated by expressing their peak area as a percentage of total hydrocarbons detected.

To summarize the overall alkane profile a weighted average carbon chain number (N) was calculated as follows:

$$N = \Sigma a_i p_i,$$

where a_i is an alkane with i carbons and a proportional concentration of p_i. Analysis of variance was used to test regional and population effects on the relative composition of the major alkanes and mean weighted alkane length (N). Species was treated as a fixed effect and geographic regions within species and populations within regions as random effects in a nested model. Variance components were estimated using the maximum likelihood method.

Correspondence analysis was carried out on a matrix of individual trees as operational taxonomic units and percentage composition of 11 alkanes (C_{23} to C_{33}) as variables, to seek multivariate pattern in the data.

To test hypotheses that relative composition of major alkanes and N were: 1. likely to be due to selection, or 2. due to divergence through isolation, Mantel tests were constructed to compare population taxonomic distance matrices of alkane composition, climatic variables and geographic distance. The geographic distance matrix that was a modification of that proposed by Douglas & Endler (1982), was derived in four stages: 1. inter-population distances from within the same region (as defined in Table 1) were actual distances in km, 2. inter-population distances between regions were assigned the highest of the within-region distances, 3. all inter-population distances were assigned a weighting (one for populations from the same coastline: African Atlantic, American Atlantic, American Pacific; two for population pairs from African and American sources), 4. a second weighting was applied to all inter-population distances (one for At-

Table 1. Sampling Locations and some climatic data

Locality	Lat/Long	Mean annual precip.	Mean maximum temp.	Mean minimum temp.	Geog. group
R. mangle					
Guadeloupe	16.3° N, 61.5° W	1790	26.0	22.8	2
Guadeloupe	16.3° N, 61.5° W	1790	26.0	22.8	2
Fr. Guiana, Cayenne	4.8° N, 52.4° W	3300	26.7	25.4	1
Fr. Guiana, Crique Patate	4.8° N, 52.4° W	3300	26.7	25.4	1
Fr. Guiana, Sinnamary	5.2° N, 52.7° W	3200	26.6	25.7	1
Dominican Republic	18.4° N, 69.8° W	1890	27.1	24.2	2
Dominican Republic	18.4° N, 69.8° W	1890	27.1	24.2	2
Dominican Republic	18.4° N, 69.8° W	1890	27.1	24.2	2
Mexico, Mazatlan	23.2° N, 105.4° W	980	28.3	19.9	3
Mexico, Tecuala	22.5° N, 105.0° W	1000	28.3	20.0	3
Florida, Naples	26.2° N, 81.8° W	1340	28.0	18.1	4
Florida, Flamingo Park	25.2° N, 80.9° W	1280	27.5	18.6	4
Florida, Keys	25.0° N, 80.5° W	1200	29.0	20.7	4
Florida, South Miami	25.7° N, 80.3° W	1500	28.2	20.2	4
Joals, Senegal	14.4° N, 17.0° W	650	27.7	20.9	5
Gabon, Port Gentil	0.7° S, 8.7° E	1950	27.3	23.3	6
R. racemosa					
Fr. Guiana, Cayenne	4.8° N, 52.4° W	3300	26.7	25.4	1
Fr. Guiana, Crique Patate	4.8° N, 52.4° W	3300	26.7	25.4	1
Fr. Guiana, Sinnamary	5.2° N, 52.7° W	3200	26.6	25.7	1
Gambia	13.5° N, 16.4° W	870	27.0	23.2	5
Senegal, Ziguinchor	12.6° N, 16.2° E	1470	28.2	23.8	5
Togo, Agbanakin	6.2° N, 1.2° E	1160	28.1	24.8	6
Togo, Agonegan	6.2° N, 1.2° E	1160	28.1	24.8	6
Gabon, Port Gentil	0.7° S, 8.7° E	1950	27.3	23.3	6
R. harrisonii					
Senegal, Missirha	14.0° N 16.5° W	760	30.0	24.8	5
Guinea, Dubreka	10.0° N 13.8° W	3800	27.5	24.9	7
Gabon, Port Gentil	0.7° S 8.7° E	1950	27.3	23.3	6

lantic populations, two for population pairs from the Pacific and Atlantic). This modified geographic distance matrix took account of the reduced likelihood of trans-Atlantic migration and the isolation of Pacific coast populations due to the Isthmus of Panama.

Results

A series of straight chain alkanes, ranging from 23 to 35 carbons were detected in all populations; in most cases, the two longest chains were present in only trace amounts. In addition, several methyl-branched hydrocarbons were detected, but these were commonly in low concentrations and were not present in all samples. Since the biophysical properties of straight chain alkane mixtures is better understood than mixtures with branched compounds, we focus here on the former, including those with carbon numbers ranging from 23 to 33.

Species effects

The dominant alkanes (relative composition greater

Table 2. Analysis of variance for 4 major alkanes and weighted mean carbon number (N). Variance components estimated by maximum likelihood method are shown in parentheses

Source	df	F-ratio					P>F (variance component%)				
		C_{27}	C_{28}	C_{29}	C_{30}	N	C_{27}	C_{28}	C_{29}	C_{30}	N
Species	2	61.4	42.8	46.3	139.5		0.0001	0.0001	0.0001	0.0001	
R. mangle.											
Region	6	11.7	23.8	42.6	27.3	19.5	0.0001(15)	0.0001(49)	0.0001(58)	0.0001(42)	0.0001(22)
Population (region)	9	6.4	2.19	5.0	4.1	7.9	0.0001(27)	0.027 (6)	0.0001(13)	0.0001(14)	0.0001(35)
Error	136						(58)	(45)	(29)	(44)	(43)
R. racemosa											
Region	3	19.1	5.0	2.4	8.1	13.7	0.0001(43)	0.004 (15)	0.07(6)	0.0001(18)	0.0001(40)
Population (region)	4	3.0	0.8	2.4	2.1	0.6	0.03 (14)	0.5 (0)	0.06 (3)	0.09 (10)	0.6 (0)
Error	55						(43)	(85)	(91)	(72)	(60)
R. harrisonii											
Population	2	2.7	9.1	11.8	3.7	7.5	0.09 (15)	0.001 (26)	0.0002(36)	0.04 (16)	0.003 (26)
Error	27						(85)	(74)	(64)	(84)	(74)
Total error	217										
Corrected Total	244										

Table 3. Mean alkane composition for three species of *Rhizophora*. Same letters in rows indicate no significant difference at Bonferroni adjusted 5% significance level

	R. mangle	*R. racemosa*	*R. harrisonii*
C_{27}	10.7a	6.5b	6.5b
C_{28}	25.0a	13.9b	11.5b
C_{29}	25.4a	33.9b	39.9c
C_{31}	15.8a	29.1b	25.5c
N	28.3a	29.4b	29.1c

than 10% in most populations) were C_{27}, C_{28}, C_{29} and C_{31}. Results of nested analyses of variance showed significant species, geographic region within species and population within region effects (Table 2) for these four alkanes. Concentrations of C_{27} and C_{28} were significantly higher (Bonferroni $P>0.05$), and levels of C_{29} and C_{31} lower in cuticular waxes of *R. mangle* (Table 3). Whereas, C_{27} and C_{28} did not differ significantly between *R. racemosa* and *R. harrisonii*, levels of C_{29} were lower, and C_{31} higher in *R. racemosa* (Table 3). These differences in concentrations of the four major alkanes were reflected in indices for the overall alkane profiles. Weighted mean carbon number (N) was 1 carbon shorter in *R. mangle* compared with *R. racemosa* (Table 3). Although the difference in N between *R. harrisonii* and *R. racemosa* was small, it was significant according to the Bonferroni $P<0.05$ test.

Correspondence analysis carried out on all alkanes from C_{23} to C_{33} yielded two eigenvectors explaining 72.5% of total inertia. These two vectors, therefore, summarized the data without much loss of information. The plot of population means (Figure 1) gave a good separation of most populations of *R. mangle* from the two other species; only populations from Mexico and Senegal showing some overlap with *R. racemosa* and *R. harrisonii*, due to a higher relative composition of long chain alkanes. New World populations of *R. mangle* separated into a Floridan group with high C_{27} and shorter chain alkanes and a Caribbean group high in the C_{28} alkane. French Guianan populations were intermediate between these two. The three populations of *R. harrisonii* were well dispersed amongst the scatter of populations of *R. racemosa*.

Within-species effects

R. mangle

The 17 sampled populations of *R. mangle* covered the broad range of this species in the Atlantic and Caribbean region, as well as Pacific coastal localities in Mexico. Within this geographic range, sites range from humid tropics as in Guiana and Gabon, through seasonal climates with varying rainfall along the Caribbean and Floridan coasts to more arid conditions in Senegal and Pacific Mexico. Mean concentrations of the four dominant alkanes are shown for the seven geographic regions in Figure 2. Regions are

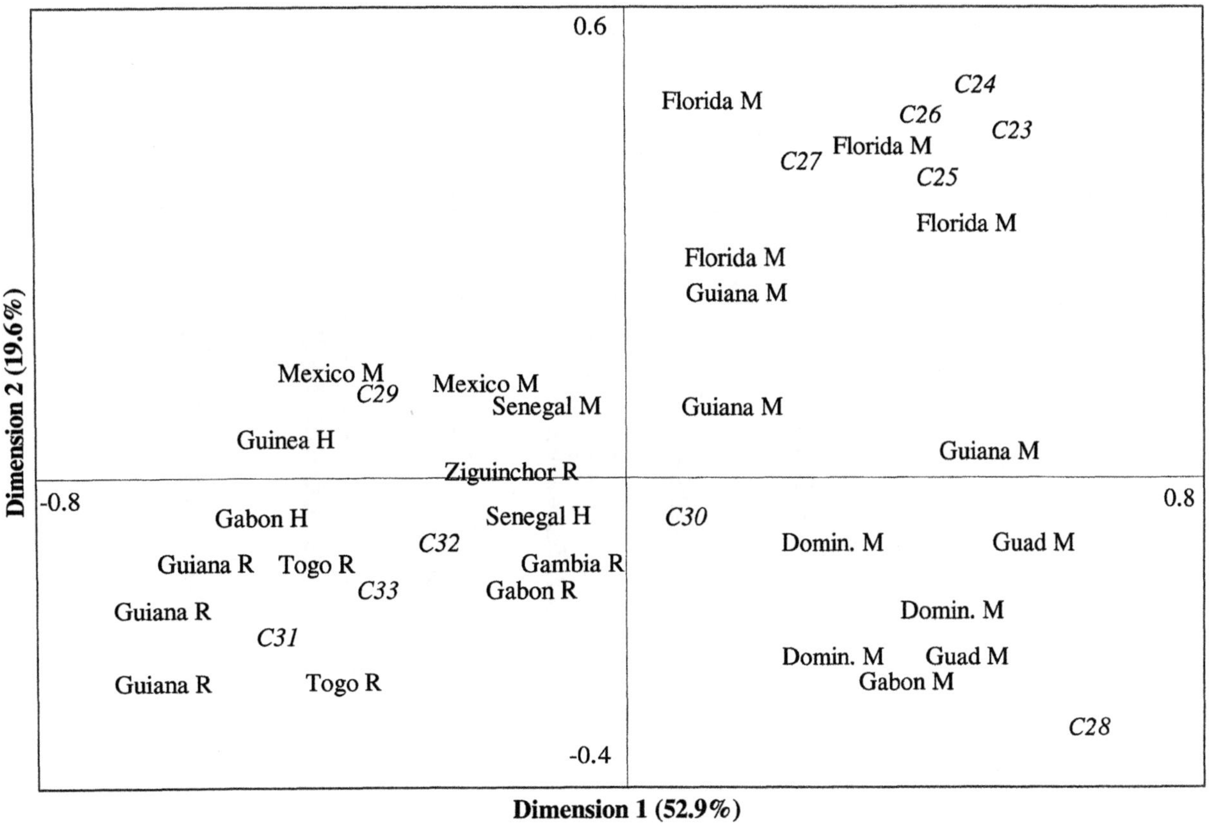

Figure 1. Correspondence analysis ordination of population means for alkane composition of cuticular waxes of three species of *Rhizophora*: Letters after population locations refer to species: H-*R. harrisonii*; M-*R. mangle*; R-*R. racemosa*. Simultaneous ordination of variables shown in italics, where numbers refer to number of carbon atoms.

ordered from left to right according to increasing annual rainfall. Although the C_{27} alkane varied among regions (Figure 2), there appeared to be no clear climatic pattern in this variation. By contrast, C_{28}, was the single most important hydrocarbon among tropical and Caribbean populations, falling to second in importance in Floridan populations and to the least important in populations from Senegal and Pacific Mexico. In a reverse trend, C_{29}, which was the most important alkane in Floridan, Mexican and Senegal populations, was of less importance in tropical and Caribbean localities. Emphasizing the trend towards the dominance of longer chain alkanes in the more arid localities, C_{31} reached highest concentrations in the Mexican and Senegal populations, and in the latter was almost equal in proportion to C_{29}.

Variance components estimates showed that geographic region was the most important contributor to overall variation in levels of C_{28} and C_{29}, and of approximately equal importance with error variation for C_{31} (Table 2). However, it contributed only 15% to the variation in levels of C_{27}. Error variation in these analyses includes within population variation in addition to sampling and measurement error, therefore genetic variation among trees could be important. As shown in Table 2, the population within region effects were significant, but the variance components estimates indicate that for the alkanes C_{28}, C_{29} and C_{31}, variation among populations within regions was a relatively minor contributor to total variation. This was not the case for N (weighted mean carbon number), for which population within region accounted for a higher proportion of variation than did geographic region. This was probably a result of N being calculated from all alkanes; variation within the minor alkanes may not be driven by the same forces as those determining variation in the dominant alkanes.

Although the analyses of variance indicated the importance of regional effects on alkane proportions,

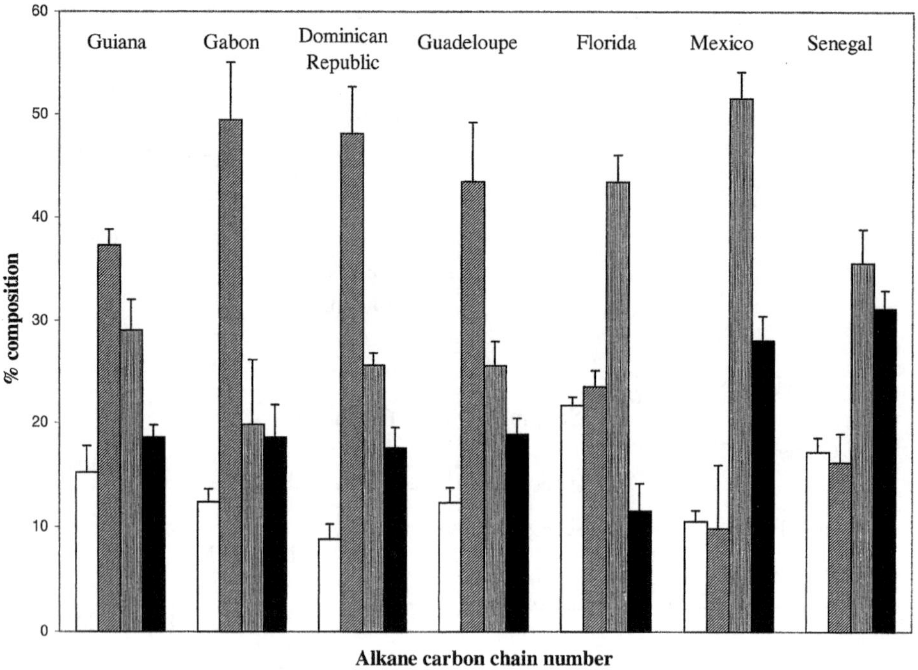

Figure 2. Histogram of mean percentage composition of the four major alkanes for geographic regions of *R. mangle*. No shading – C_{27}; diagonal shading – C_{28}; horizontal shading – C_{29}; full shading – C_{31}. Length of vertical bars is 2 standard errors shown on one side of the histogram.

further tests are needed to try to separate effects of isolation by distance from effects of selection pressure from climatic conditions. Mantel tests were used to compare taxonomic distance among all pairs of populations for each of the four dominant alkanes and mean weighted alkane number (N) with taxonomic distances for the three climatic variables and a weighted geographic distance. The Mantel tests showed no significant covariation between any of the alkane distance matrices and the geographic distance matrix. Both the C_{28} and C_{29} distance matrices covaried with mean maximum temperature and C_{28} and C_{31} distance matrices covaried with annual rainfall (Table 4).

R. racemosa

To carry out a full nested analysis of variance, as was the case for *R. mangle*, populations from Ziguinchor and Gambia were combined under a single geographic region. Although a fairly strong climatic gradient of decreasing rainfall with latitude exits within this area, populations are in moderately close geographic proximity. Analysis of variance indicated significant effects of geographic region (Table 2), but there was no trend of increasing proportion of long chain alkanes in the drier localities, as was the case for *R. mangle*. Indeed, levels of C_{27} were highest and C_{29} and C_{31} among the lowest in the more arid North-West African populations (Figure 3). Alkane profiles in which longer carbon chains dominated were characteristic of Guianan populations. Variance components estimates show that, except for C_{27}, geographic region and population within region have relatively modest effects on total variation in the alkane proportions, most of the variation being in the error term (Table 2). This suggests that tree-to-tree variation may be much more important in *R. racemosa* than it is in *R. mangle*.

From Mantel tests, the C_{28} distance matrix covaried significantly with geographic distance, annual rainfall and mean minimum temperature. Distance matrices for C_{31} and for mean weighted carbon number (N) covaried with mean minimum temperature (Table 4).

R. harrisonii

Sampling of *R. harrisonii* was more limited, with populations only available from West Africa. A nested analysis of variance design was not possible, so that only population effects are considered here (Table 2). As for *R. racemosa*, any differences among populations tended towards a decrease in the relative proportions of the longer chain alkanes in the more arid

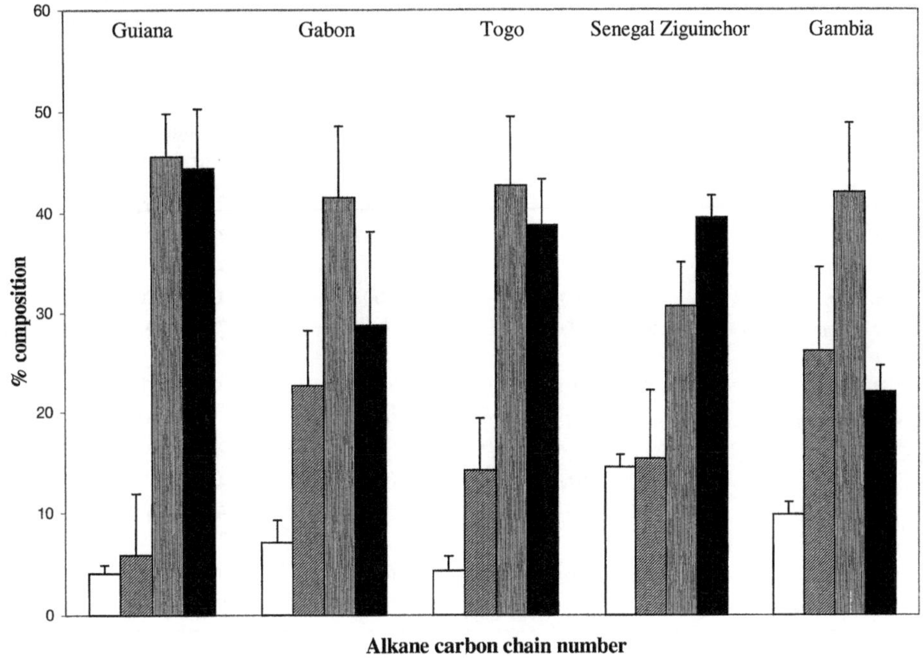

Figure 3. Histogram of mean percentage composition of the four major alkanes for geographic regions of *R. racemosa*. No shading – C_{27}; diagonal shading – C_{28}; horizontal shading – C_{29}; full shading – C_{31}. Length of vertical bars is 2 standard errors shown on one side of the histogram.

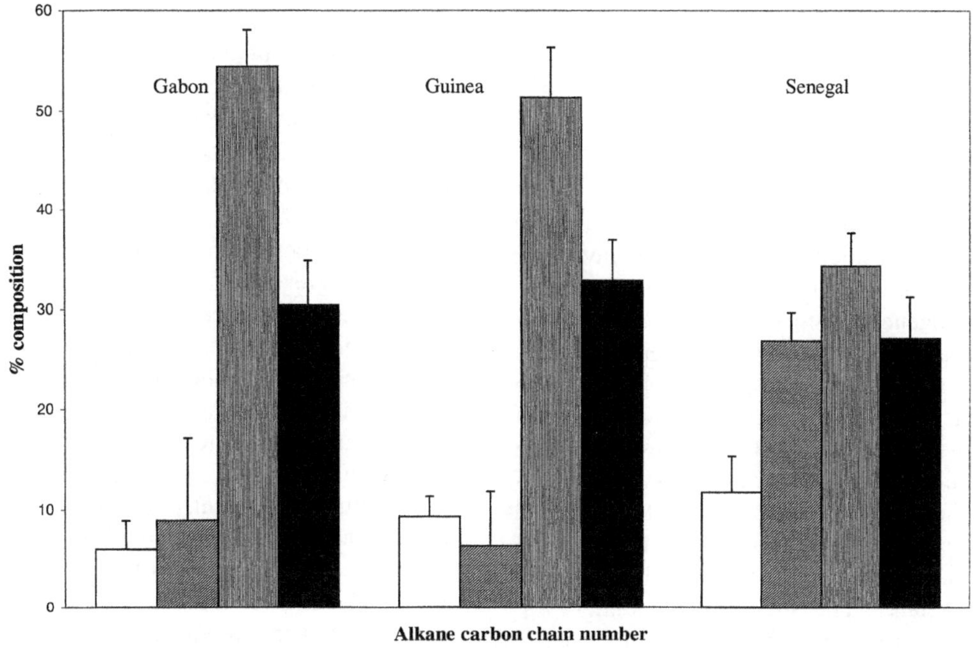

Figure 4. Histogram of mean percentage composition of the four major alkanes for populations of *R. harrisonii*. No shading – C_{27}; diagonal shading – C_{28}; horizontal shading – C_{29}; full shading – C_{31}. Length of vertical bars is 2 standard errors shown on one side of the histogram.

Table 4. Pair-wise comparisons of taxonomic distance matrices for alkane data with climatic and geographic distance matrices. Values of Mantel Z statistic, with probabilities that random Z (based on 1000 permutations) exceeds observed Z shown in parentheses, in bold where they exceed the Bonferroni 5% critical level ($p < 0.0025$)

	Species	annual rainfall	Mean maximum temperature	Mean minimum temperature	Geographic region
C_{27}	*R. mangle*	−0.11 (0.74)	−0.09 (0.80)	0.28 (0.02)	0.001 (0.50)
	R. racemosa	−0.047 (0.62)	0.009 (0.57)	0.35 (0.03)	−0.08 (0.69)
C_{28}	*R. mangle*	**0.38 (0.002)**	**0.42 (0.002)**	0.12 (0.10)	0.19 (0.06)
	R. racemosa	**0.61 (0.0001)**	0.20 (0.14)	**0.79 (0.0001)**	**0.57 (0.001)**
C_{29}	*R. mangle*	0.13 (0.09)	**0.50 (0.001)**	0.13 (0.10)	0.25 (0.05)
	R. racemosa	0.007 (0.49)	0.20 (0.11)	0.15 (0.21)	0.07 (0.36)
C_{31}	*R. mangle*	**0.67 (0.001)**	0.02 (0.39)	0.20 (0.06)	−0.05 (0.37)
	R. racemosa	0.28 (0.05)	−0.08 (0.69)	**0.69 (0.0001)**	0.23 (0.10)
N	*R. mangle*	−0.09 (0.26)	0.04 (0.34)	0.13 (0.18)	−0.002 (0.47)
	R. racemosa	0.33 (0.05)	0.004 (0.49)	**0.91 (0.0001)**	0.3 (0.06)

north-west African population (Figure 4). Among the three populations studied here, it is interesting to note that the alkane profile from Senegal contrasted sharply with that from Guinea and Gabon.

Discussion

The evolution of patterns of differentiation in cuticular hydrocarbon composition may be complex, involving effects of genetic adaptation to environmental conditions, phylogenetic history and genetic drift. Our results indicate that environmental adaptation may be an important factor in determining hydrocarbon variation, at least in *R. mangle*, which occupies the greatest range of environmental conditions of the three species studied here. Mantel tests showed significant correlations between taxonomic distances based on the major alkanes, and taxonomic distances of climatic variables, but no correlation with weighted geographic distance. This suggests that parallel evolution for these phenotypic characters occurs even in populations that have been isolated for a long period of time, as is the case for Pacific coastal populations from Mexico. The proportion of variance associated with geographic region (where regions are considered more or less climatically homogeneous) was much lower in *R. racemosa* than in *R. mangle*.

If natural selection was a potent force in differentiation in alkane composition, a similar response might be expected from broadly sympatric species. That this was not the case seems to underline the different habitat preferences of the two species that have been observed from field studies (Savory, 1953; Jonker, 1959; Bretteler, 1977). The preference of *R. racemosa* for lower salinity to freshwater conditions may be a key factor in diminishing the selective advantage of cuticular wax composition. For this species, Mantel tests revealed a significant correlation between the population distance matrix for the long chain C_{31} alkane and the distance matrix for mean minimum temperature. Both climatic effects (minimum temperature and annual rainfall) and geographic distance were significantly correlated with the population distance matrix for the C_{28} alkane. Care is needed in interpreting these results for the C_{28} alkane, since the distance matrix for rainfall and geographic distance are confounded; both having maximum values for population pairs that include French Guiana. This is an unfortunate consequence of insufficient sampling of this species from different regions within the New World. The significant Mantel test statistic between minimum temperature and taxonomic distance for two of the alkanes, suggests that minimum temperature may be an important constraint on the latitudinal range of *R. racemosa*.

Mantel test statistics are useful for seeking correlations between degree of divergence in a phenotypic character and degree of divergence in environmental variables. However, the Mantel statistic does not reveal the direction in which a phenotypic variable is correlated with an environmental variable. In *R. mangle*, concentrations of longer chain alkanes were higher in more arid conditions. Thus, C_{29} and C_{31} take on greater importance in the cuticular waxes of populations from Senegal, Mexico and, to a lesser extent, Florida. This pattern of increasing alkane chain length under more arid conditions is consistent with the expectations of biophysical properties of waxes. Transition melting temperatures of alkanes increase with increasing number of carbon atoms (CRC, 1992) and alkane mixtures broaden the range of temperature transition (Gibbs, 1995). Reiderer & Schneider (1990) suggest that increases in wax crystallinity (reduced permeability) are associated with increases in the ratio of average weighted carbon number (N in our data) to dispersion about this average number. Therefore, the alkane composition of cuticular waxes from Senegal and Mexican populations of *R. mangle* can be expected to contribute to reduced cuticular permeability. Interestingly, this pattern of population variation in alkane composition was not observed in *R. racemosa* or in *R. harrisonii*. Indeed, the tendency in these two species was for longer chain alkanes to be more important in populations from tropical regions. As mentioned earlier, selection pressure may not be the same in these three species because of different habitat preferences.

Assuming selection pressure to be an important factor in differentiation of cuticular waxes, species from more arid zones should, on average, have a greater abundance of longer chain alkanes. Our data seem to be in contradiction with this, since *R. mangle* had the lowest mean weighted carbon number. However, speciation appears to commonly be associated with more radical changes in alkane profiles than intraspecific differentiation. Presumably, shifts in biosynthetic pathways may occur in relation to environmental conditions prevailing at the time and location of the speciation event. The potential for subsequent intraspecific variation may then vary among species. Clearly, phylogenetic history may also confound any interspecific comparisons. A better understanding of the phylogeny of this group and likely geographic origins will help us better understand the species differences. As a tentative hypothesis, we propose that *R. mangle* is ancestral to the two other Atlantic species, and that its origin was in warm, humid coastal sites. Under these environmental conditions, there would have been no strong force favouring long chain alkanes. It will be interesting to compare Pacific members of the genus, that may help shed light on possible ancestral migration patterns.

Acknowledgements

The authors thank A. Bousquet-Mélou, D. Imbert and F. Fromard for providing some of the plant material and M.-C. Souqual for technical assistance.

References

A.-Rafii, Z., R. S. Dodd & F. Fromard, 1996. Biogeographic variation in foliar waxes of mangrove species. Biochem. Syst. Ecol. 24: 341–345.

Breteler, F. J., 1969. The Atlantic species of *Rhizophora*. Acta Bot. Neerl. 18: 434–441.

Breteler, F. J., 1977. America's Pacific species of *Rhizophora*. Acta Bot. Neerl. 26: 225–230.

CRC, 1992. CRC Handbook of chemistry and physics. CRC Press, Boca Raton, Florida.

Dodd, R. S., Z. A.-Rafii & A. B. Power, 1998. Ecotypic adaptation in *Austrocedrus chilensis* in cuticular hydrocarbon composition. New Phytol. 138: 699–708.

Dodd, R. S., F. Blasco, Z. A.-Rafii & E. Torquebiau, 1999. Mangroves of the United Arab Emirates: ecotypic diversity in cuticular waxes at the bioclimatic extreme. Aquat. Bot. 63: 291–304.

Douglas, M. E. & J. A. Endler, 1982. Quantitative matrix comparisons in ecological and evolutionary investigations. J. theor. Biol. 99: 777–795.

Gentry, A. H., 1982. Phytogeographic patterns as evidence for a Choco refuge. In Prance, G. T. (ed.), Biological Diversification in the Tropics. Columbia University Press, New York: 112–136.

Gibbs, A. G., 1995. Physical properties of insect cuticular hydrocarbons: model mixtures and lipid interaction. Comp. Biochem. Physiol. 112B: 667–672.

Jiménez, J. A., 1987. A clarification on the existence of *Rhizophora* species along the Pacific coast of Central America. Brenesia 28: 25–32.

Lockey, K. H., 1988. Lipids of the insect cuticle: origin, composition and function. Comp. Biochem. Physiol. 89B: 595–645.

Jonker, F. P., 1959. The genus *Rhizophora* in Suriname. Acta Bot. Neerl. 8: 58–60.

Keay, R. W. J., 1953. *Rhizophora* in West Africa. Kew Bull. 1: 121–127.

Prance, G. T., M. Freitas da Silva, B. W. Albuquerque, I. De J. da Silva Araújo, L. M. Medeiros Carreira, M. M. Nogueira Braga, M. Macedo, P. N. da Conceiçao, P. L. Braga Lisbôa, P. I. Braga, R. C. Lobata Lisbôa & R. C. Quieroz Vilhena, 1975. Revisao taxônomicadas espécies amazônica de Rhizophoraceae. Acta Amazonica 5: 5–22.

Reiderer, M. & G. Schneider, 1990. The effect of the environment on the permeability and composition of *Citrus* leaf cuticles. Planta 180: 154–165.

Savory, H. J., 1953. A note on the ecology of *Rhizophora* in Nigeria. Kew Bull. 1: 127–128.

Hydrobiologia **413:** 11–16, 1999.
R.S. Dodd (ed.), Diversity and Function in Mangrove Ecosystems.

Genetic differentiation between *Bruguiera gymnorhiza* and *B. sexangula* in Sri Lanka

Pushpa Damayanthi Abeysinghe[1,2], Ludwig Triest[1,*], Bart De Greef[1], Nico Koedam[1] & Sanath Hettiarachi[2]
[1]*Vrije Universiteit Brussel, Research Group General Botany and Nature Management, Pleinlaan 2, B-1050, Brussel, Belgium. E-mail: ltriest@vub.ac.be*
[2]*Dept. of Botany, University of Ruhuna, Matara, Sri Lanka*
*(*Author for correspondence)*

Key words: DNA, RAPD, genetic diversity, *Bruguiera*, Sri Lanka, mangroves

Abstract

The identification of populations of *Bruguiera sexangula*, *Bruguiera gymnorhiza* and their putative hybrids in the field is difficult using only morphological and phenological characters. Using a PCR based technique, RAPD (Random Amplified Polymorphic DNA), the genetic variation of *Bruguiera* populations was studied from contrasting climatic and geographic regions along the southwest coastal region of Sri Lanka. Out of 45 primers screened, 20 primers allowed us to observe polymorphism, not only between species (interspecific) but also within the species (intraspecific). Analysis of RAPD data appears to be helpful in determining the genetic relationship among populations of *B. gymnorhiza* and *B. sexangula*. RAPD markers revealed that the two species are well separated without any hybrid position between the two taxa though they occur in mixed stands. Although sampling sizes of populations of this study were small, genetic variation among *B. gymnorhiza* and *B. sexangula* populations could be observed. For *B. sexangula*, it was possible to differentiate each of the three populations, even when using a small number of primers.

Introduction

Mangroves or mangals are tropical and subtropical forests developed in the intertidal zone (Arulchelvam, 1968; Lindén & Jernelöv, 1980; De Silva & Balasubramaniam, 1985; Tomlinson, 1986; Amarasinghe & Balasubramaniam, 1992; McKee, 1995; Pemadasa, 1996). Though mangroves occur in tropical and subtropical areas of the world, the most luxuriant mangrove forests are found in Southeast Asia (Lindén & Jernelöv, 1980). The present extent of mangroves in Sri Lanka has been estimated at 4000 ha (Pemadasa, 1996). Sri Lankan mangroves include approximately 23 mangrove species (Jayawardene, 1985) from nine genera (*Rhizophora, Bruguiera, Ceriops, Avicennia, Xylocarpus, Excoecaria, Lumnitzera, Aegiceras* and *Nypa*: definition according to Tomlinson, 1986).

Mangroves are ecologically important components of the coastal ecosystems because they have developed on sheltered shores of lagoons and estuaries, and because they prevent sea erosion and act as a land building system. Physiologically and morphologically, they have various features for extremely harsh and hostile dry atmospheric conditions and wet, anoxic and saline root environments.

Economically, the mangroves are providing suitable habitats for crustaceans and fish, and they deliver plant products such as tannins and dyes, timber, resources for alcoholic beverages and food (Lindén & Jernelöv, 1980; De Silva & Balasubramaniam, 1985; Jayawardene, 1985; Tomlinson, 1986; Pemadasa, 1996; Hellier, 1988). As a result of human activities, extensive mangrove areas are being lost at an alarming rate (Lindén & Jernelöv, 1980; Jayawardene, 1985; Hellier, 1988; Pemadasa, 1996).

Despite the great interest in their conservation, in management of mangrove species (Jayawardene, 1985; Dodd et al., 1995; Farnsworth & Ellison, 1997)

and in their intrinsic value as a genetic resource, relatively few genetic studies have been conducted. Some studies have been reported on the genetics of flowering time and morphological characters (Duke, 1984; Duke, 1995). Isozymes (McMillan, 1986; Goodall & Stoddart, 1989) have been used rarely for a systematic assessment of genetic variability in mangroves. Albino genotypes occurred by either nuclear or plastid genome mutations of *Rhizophora mangle* L. and have been fingerprinted (Corredor et al., 1995; Klekowski et al., 1996). Variation among *Rhizophora* had been studied using epicuticular wax composition, hydrocarbons and triterpenoids (Dodd et al., 1995), using RAPD (Random Amplified Polymorphic DNA) and restriction fragment length polymorphism (Parani et al., 1997b). Traditional measurements of genetic diversity of plants rely upon the ability to differentiate morphological, physiological or biochemical characters (Khasa & Dancik, 1996). Though *Bruguiera* species can usually be distinguished by morphological (Arulchelvam, 1968) and phenological traits (P. L. Hettiarachchi, pers. com. University of Sri Jayawadenepura, Sri Lanka), most of the morphological characters used as descriptors can only be assessed at maturity. Particular phenological traits can only be measured at certain times of the year, while other morphological traits require extensive observation during the whole growing season of the plant. In the field, hybridisation and introgression (putative hybrids or intermediates) between *B. sexangula* and *B. gymnorhiza* can be observed, when using morphological criteria.

RAPD has been shown to be a very useful technique for the construction of genetic maps (Tingey & Del Tufo, 1993), DNA fingerprinting (Welsh & McClelland, 1990; Williams et al., 1990), the identification of genetic polymorphism and parentage determination (Welsh et al., 1991).

Despite the great interest in management and conservation of tropical tree species, a very limited number of researches have been conducted on the population genetics of tropical tree species (Dawson et al., 1996; Parani et al., 1997a,b; Schierenbeck et al., 1997). The identification of a species, information on the geographic structure and precise understanding of the organisation of the existing genetic diversity within and among populations are very important for the conservation of genetic resources (Triest, 1991; Aldrich et al., 1992; Russell et al., 1993; Dawson et al., 1995, 1996). For management and conservation purposes, the identification of *B. sexangula* and *B. gymnorhiza* and their putative hybrids or introgressed intermediates, even at the very young stage of their life cycle e.g. seedling stage, could be accomplished using RAPD markers.

The objective of this paper was to evaluate the use of RAPD to genetically differentiate the two sympatric *Bruguiera* species which originate from different geographic and climatic areas in Sri Lanka. Consistent DNA amplification products should be selected to differentiate *B. sexangula* and *B. gymnorhiza* and assess the possible occurrence of hybrids or intermediates.

Materials and methods

Sample collection

Leaf material was collected from populations of *B. sexangula* and *B. gymnorhiza* (nomenclature according to Tomlinson, 1986 except for *B. gymnorhiza*) in order to cover the contrasting climatic zones (wet zone, dry zone and intermediate zone) and different geographical areas of the island (Figure 1). A total of six localities were chosen from the western to the southeastern coastal region of Sri Lanka (Table 1). Climatically the intermediate zone covers the sampling sites in Pambala (P) and Rekawa (R, disjunct along the coast) which are well separated geographic locations in the intermediate zone of the island. Merawala (M) is located in the dry zone of the island. Dondra (D) and Kudawella (K) are geographically close in the wet zone, while Negombo (N) is at the limit of the wet zone (Müller-Dombois, 1968). These locations were selected in order to resolve the maximal geographic variation on the island. During collection, morphological criteria were applied to classify populations as either *B. sexangula* or *B. gymnorhiza* (Arulchelvam, 1968).

DNA extraction and amplification

Total genomic plant DNA was extracted from young frozen leaf samples. Leaf tissue was ground to a powder in liquid nitrogen using a mortar and pestle, and then 5 ml of extraction buffer was added. The extraction buffer consisted of 3% of CTAB (hexadecyl trimethyl ammonium bromide), 2 M NaCl, 100 mM Tris–HCl pH 8.0, 1% β-mercaptoethanol, 20 mM EDTA and 1% PVP (polyvinyl polypyrrolidone). The homogenate was incubated at 65 °C for 30 min, 5 ml of chloroform/isoamyl alcohol (24:1) was added and the mixture was centrifuged at 3000 g for 10 min. The supernatant was transferred to a new 50 ml

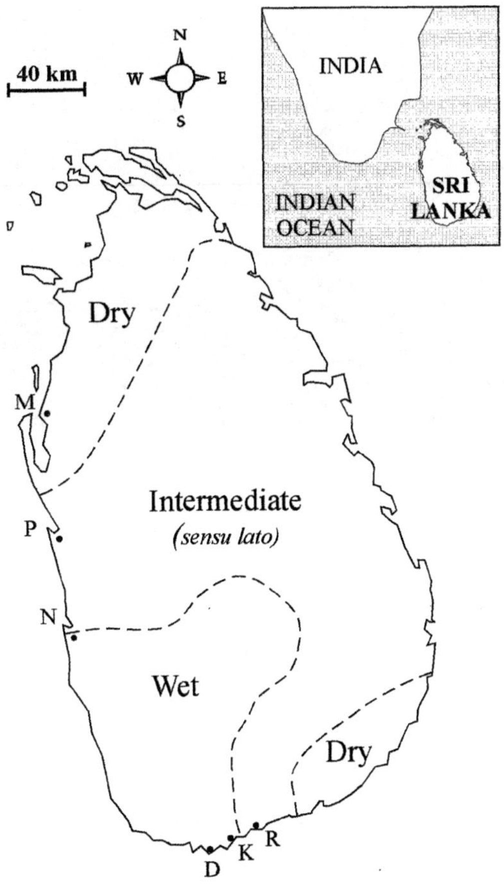

Figure 1. Geographical distribution of sampling locations for *B. gymnorhiza* and *B. sexangula* in Sri Lanka: Merawala (M), Pambala (P), Negombo (N), Dondra (D), Kudawella (K) and Rekawa (R). Adapted from Dahdouh-Guebas et al.

Table 1. The location, climatic zones and number of individuals of *B. gymnorhiza* and *B. sexangula* investigated for RAPD (climatic zones according to Müller-Dombois, 1968)

Species		Climatic zone	No. of individuals
B. gymnorhiza	Merawala (M)	Dry	02
	Negombo (N)	Wet	10
	Pambala (P)	Intermediate	01
	Rekawa (R)	Intermediate	04
B. sexangula	Rekawa (R)	Intermediate	06
	Kudawella (K)	Wet	06
	Dondra (D)	Wet	04

centrifuge tube (Greiner) which was filled with chloroform/isoamyl alcohol (24:1) and centrifuged again for 10 min. Once again, the supernatant was transferred to a new 50 ml tube. DNA was precipitated by mixing with 0.6 volume ice-cold isopropanol, then kept at room temperature for 15 min and pelleted by centrifugation at 7700 g for 15 min. The supernatant was removed, washed with 70% ethanol to get rid of salts, air dried and dissolved in sterile water. This was treated with 10 mg/ml RNase A (Sigma). The quality of the DNA was monitored by scanning with a UV spectrophotometer from 200 nm to 300 nm. The concentration of the DNA samples was estimated by gel electrophoresis against a standard DNA (lambda-DNA digested with Pst 1). Amplification was carried out according to Williams et al. (1990) and amplification products were size-fractionated on a 1% agarose gel containing 10 μg/ml of ethidium bromide. They were visualized under UV light and photographed for further analysis.

Forty five decanucleotide primers were screened using the template DNA from eight individuals from *B. sexangula* and *B. gymnorhiza* to identify informative primers that generated polymorphic patterns of PCR (Polymerase Chain Reaction) products. Fragments were considered polymorphic when both absence and presence of the same fragment in the same species was observed and monomorphic when a fragment was present among all individuals within the species. Reproducibility of the amplified products was tested repeating the same reaction at least twice. No amplification was found in the control sample in which no DNA was added to the reaction mixture indicating that all the fragments observed in other samples are not artefacts, but due to the amplification of genomic DNA (Williams et al., 1990). A similarity matrix using the simple matching coefficient and a Principal Coordinate analysis (PCOORDA) were performed using NTSYS-pc (Rohlf, 1993).

Results and discussion

Primer screening and discrimination of B. sexangula *and* B. gymnorhiza

The sizes of the amplification products were 0.5–2.8 kb. According to the patterns of the amplified products, primers were categorized into three groups which were interspecific, intraspecific and both (Table 2). Out of 45 primers, 20 informative primers

Table 2. The number of primers used for amplification of genomic DNA from *B. gymnorhiza* (Bg) or *B. sexangula* (Bs) and the number of amplified products (loci) generated for each primer

Primer	Sequence	No of polymorphic band		
		Bg	Bs	Both
OPA-14	TCTGTGCTGG	01	–	02
OPA-15	TTCCGAACCC	01	–	01
OPE-07	AGATGCAGCC	02	–	02
OPE-14	TGCGGCTGAG	–	–01	
OPE-15	ACGCACAACC	02	–	01
OPE-16	GGTGACTGTG	01	02	–
OPJ-07	CCTCTCGACA	–	0101	
OPJ-11	ACTCTCGACA	01	0101	
OPJ-12	GGGTGTGTAG	–	–	02
OPJ-20	AAGCGGCCTC	–	0102	
OP0-15	TGGCGTCCTT	02	—	
OPT-06	CAAGGGCAGA	01	0101	
Total	12	11	06	14

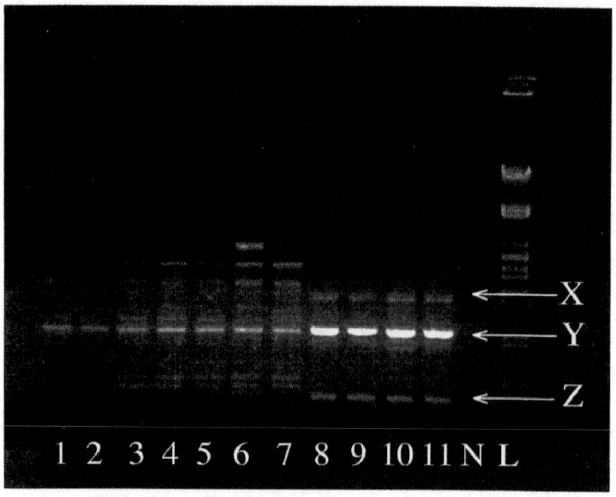

Figure 2. PCR amplification of genomic DNA from *B. gymnorhiza* and *B. sexangula*, with primer OPE-16 showing 3 RAPD markers. DNA. Lane 1–7 is *B. gymnorhiza*, lane 8–11 is *B. sexangula*, and lane 13 is the molecular weight marker (L-Lambda DNA digested with Pst I). The negative control (lane 12) contained all reagents except template. The 0.9 kb fragment is present in both *B. gymnorhiza* and *B. sexangula* while 1.6 kb and 0.7 kb fragments are present only in *B. sexangula*.

revealed simple, unique, easily resolved bands and one or more RAPD markers. Six primers showed amplification without detection of polymorphism while two primers gave a smear pattern. Eleven primers produced complex banding patterns that were difficult to interpret and were eliminated from the analysis. The remainder (six primers) failed to amplify. The individuals showed the same amplification pattern throughout the repeats.

In this study, with 12 primers, 31 RAPD bands (loci) were obtained for both *B. gymnorhiza* and *B. sexangula*. Out of 31 markers, 14 marker alleles were present in both species (Table 2). As an example, the decanucleotide primer OPE-16 primed the amplification of eight and three fragments of *B. gymnorhiza* and *B. sexangula*, respectively. One fragment, (0.9 kb) was present in both *B. gymnorhiza* and *B. sexangula*, while the amplified fragments of 1.6 kb and 0.7 kb could only be observed in samples of *B. sexangula* (Figure 2). Amplification with primer OPO-15 showed a diagnostic and unique banding pattern for *B. gymnorhiza* and *B. sexangula* also. These results indicated that RAPD markers allow us to differentiate *B. gymnorhiza* and *B. sexangula* unambiguously and provided a useful technique for studying genetic variation among populations of *B. gymnorhiza* and *B. sexangula*.

Genetic differentiation of B. gymnorhiza *and* B. sexangula

In order to assess the inter and intraspecific variation of *Bruguiera* populations, all 31 diagnostic RAPD markers were analyzed. The computation of the simple matching coefficient using 31 RAPD loci and subsequently PCOORDA revealed variation between the two species, and without exception *B. sexangula* and *B. gymnorhiza* were separated into two distinct groups along the first coordinate axis which represents 65% of the total variation (Figure 3). Though populations displaying morphological characters of both species were observed, no genetic hybrids or intermediates of *B. sexangula* and *B. gymnorhiza* could be observed in this study, indicating that these sympatric species must have a separate gene pool. RAPD analysis and tentative morphological identifications are correlated, however an evaluation of intra and interspecific variations of *Bruguiera* populations in the field is difficult and morphological descriptors may have a utility in describing complex genetic relationships.

The genetic similarity values for *B. sexangula* ranging from about 69% to 92% indicates a possible narrow genetic base for *B. sexangula*. Multivariate analysis on RAPD markers separated the *B. sexangula* populations along the second coordinate axis into groups according to their different geographical

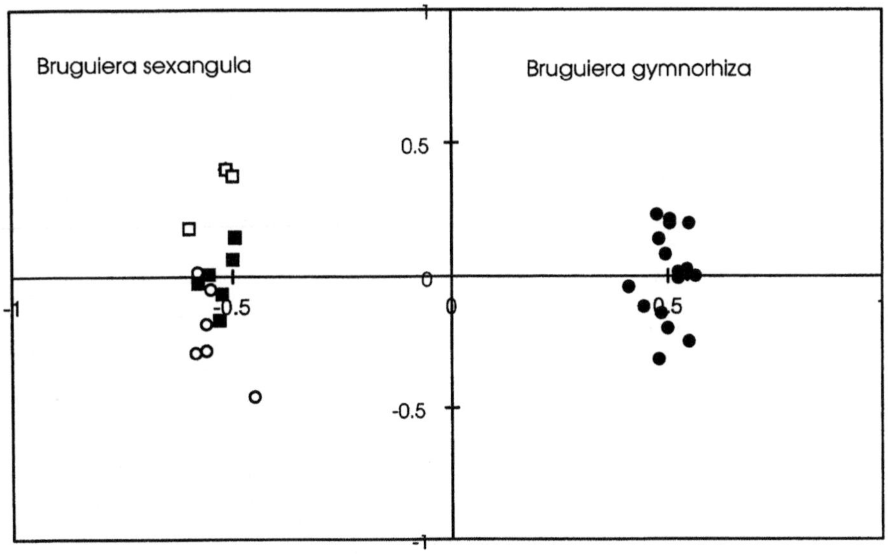

Figure 3. Principal coordinate analysis of first two components based on 31 RAPD markers for *B. gymnorhiza* and *B. sexangula* from Sri Lanka. Each symbol represents one individual. *B. sexangula*: open squares, Dondra; solid squares, Rekawa and open circle, Kudawella and solid circles, all *B. gymnorhiza* individuals.

regions. Two *B. sexangula* individuals from Dondra (small population) were identical. Genetic variation within each of the populations of *Bruguiera* may vary with the population size. Therefore, we propose to compare the small forest area at Dondra with the large population area of Negombo.

Although the present study was an initial geographical survey at species level, RAPD markers indicated possible variability among and within populations of *B. sexangula* and *B. gymnorhiza* e.g. 10 different genotypes were observed among the 10 Negombo samples which originated from the wet climatic zone. The major differences between populations may reflect the geographic differences. A next working hypothesis, therefore, would be to check whether genetic differentiation among populations of *Bruguiera* species are connected to the geographical regions that differ in their climatic regions.

Our results showed that geographical variation of populations of *B. sexangula* which originated from different locations along a coastal belt of Sri Lanka could be observed using only a few random primers. Unfortunately, there are no reports on *Bruguiera* species to compare genetic variation with our results. However, based on isozyme studies on *Avicennia germinans* (L) L. (McMillan, 1986), genetic variation has been observed on large scale geographic areas, while for an Australian mangrove, *Rhizophora stylosa* Griff., little isozyme variation had been observed. Studies with RFLP (Restriction Fragment Length Polymorphism) and RAPD of different *Avicennia* species originating from India revealed low genetic variation among populations (Parani et al., 1997a).

In conclusion, this study demonstrated that decanucleotide DNA fingerprinting was an effective method to select markers at species level and to assess the possibility of hybridisation among species. An additional working hypothesis is to get a better picture of genetic diversity within the genus based on the size of the populations with a broader array of samples and based on their origin in different climatic regions.

Acknowledgements

This research was partially funded by the ABOS (Belgian Agency for development cooperation), the Fund for Scientific Research (mandate of Research director L. Triest and FWO contract No. S2/5-ID. E53 on ecogenetic diversity in plant populations) and by the Vrije Universiteit Brussel (OZR-funding). We would like to thank the late Prof. M. A. Pemadasa, Dr L. P. Jayathissa, Kamal Jayakodi (field officer at Puttalam), Bandula Weerakonda Arachchi, Niel Thusira Kumara and R. M. Gunawardene for their painstaking assistance during the field collection in Sri Lanka.

References

Aldrich, P. R., J. Doebley, K. F. Schertz & A. Stec, 1992. Patterns of allozyme variation in cultivated and wild *Sorghum bicolor*. Theor. Appl. Genet. 8: 451–460.

Amarasinghe, M. D. & S. Balasubramaniam, 1992. Structural properties of mangrove stands on the northwestern coast of Sri Lanka. Hydrobiologia 247: 17–27.

Arulchelvam, K., 1968. Mangroves. The Ceylon Forester 314: 59–92.

Corredor, J. E., J. M. Morell, E. J. Klekowski Jr. & R. Lowenfeld, 1995. Mangrove genetics. III. Pigment fingerprinting of chlorophyll-deficient mutants. Int. J. Plant. Sci. 156: 55–60.

Dahdouh-Guebas, F., A. Verheyden, W. De Genst, S. Hettiarachi & N. Koedam, Four decade vegetation dynamics in Sri Lankan mangroves as detected from sequential aerial photography: a case study in Galle (Submitted).

Dawson, I. K., A. J. Simons, R. Waugh & W. Powell, 1995. Diversity and genetic differentiation among subpopulations of *Gliricidia sepium* revealed by PCR-based assays. Heredity 74: 10–18.

Dawson, I. K., A. J. Simons, R. Waugh & W. Powell, 1996. Detection and pattern of interspecific hybridisation between *Gliricidia sepium* and *G. maculata* in Meso-America revealed by PCR-based assays. Molec. Ecol. 5: 89–98.

De Silva, K. H. G. M. & S. Balasubramaniam, 1985. Some ecological aspects of the mangroves on the west coast of Sri Lanka. Ceylon J. Sci. 18: 22–37.

Dodd, R. S., F. Fromard, Z. A. Rafii & F. Blasco, 1995. Biodiversity among west African *Rhizophora*: Foliar Wax chemistry. Biochem. System. Ecol. 23: 859–868.

Duke, N. C., 1984. A mangrove hybrid, *Sonneratia* X *gulngai* (Sonneratiaceae) from north-eastern Australia. Austrobaileya 2: 103–105.

Duke, N. C., 1995. Genetic diversity, distributional barriers and rafting continents – more thoughts on the evolution of mangroves. Hydrobiologia 295: 167–181.

Farnsworth, E. J. & A. M. Ellison, 1997. The global conservation status of mangroves. Ambio 26: 328–334.

Goodall, J. A. & J. A. Stoddart, 1989. Techniques for the electrophoresis of mangrove tissue. Aquat. Bot. 35: 197–207.

Hellier, C., 1988. The mangrove wastelands. The Ecologist 18: 77–79.

Jayawardene, R. P., 1985. The mangroves of Sri Lanka. Bakawan 4: 8–10.

Khasa, P. D. & B. P. Dancik, 1996. Rapid identification of white-Engelmann spruce species by RAPD markers. Theor. Appl. Genet. 92: 46–52.

Klekowski, E. J. Jr., R. Lowenfeld & E. H. Klekowski, 1996. Mangrove genetics. IV. Postzygotic mutations as periclinal chimeras. Int. J. Plant. Sci. 157: 398–405.

Lindén, O. & A. Jernelöv, 1980. The mangrove swamp – an ecosystem in danger. Ambio 9: 81–88.

McKee, K. L., 1995. Interspecific variation in growth, biomass, partitioning and defensive characteristic of neotropical mangrove seedlings: response to light and nutrient availability. Am. J. Bot. 82: 299–307.

McMillan, C., 1986. Isozyme patterns among populations of black mangrove *Avicennia germinans*, from the gulf of Mexico – Caribbean and Pacific Panama. Contribution in Mar. Sci. 29: 17–25.

Müller-Dombois, D., 1968. Ecogeographical analysis of a climatic map of Ceylon with particular reference to vegetation. The Ceylon Forester 3: 144–164.

Parani, M., M. Lakshmi, S. Elango, N. Ram, C. S. Anuratha & A. Parida, 1997a. Molecular phylogeny of mangroves II. Intra- and inter-specific variation in *Avicennia* revealed by RAPD and RFLP markers. Genome 40: 487–495.

Parani, M., C. S. Rao, N. Mathan, C. S. Anuratha, K. K. & A. Parida, 1997b. Molecular phylogeny of mangroves III. Parentage analysis of a *Rhizophora* hybrid using random amplified polymorphic DNA and restriction fragment length polymorphism markers. Aquat. Bot. 58: 165–172.

Pemadasa, M., 1996. The green mantle of Sri Lanka. Tharnjee Printers, Maharagama: 1–108.

Rohlf, F. J., 1993. NTSYS-pc. Numerical taxonomy and multivariate analysis system. Applied Biostatistics Inc. New York.

Russell, J. R., F. Hosein, E. Johnson, R. Waugh & W. Powell, 1993. Genetic differentiation of cocoa (*Theobroma cacao* L.) populations revealed by RAPD analysis. Molec. Ecol. 2: 89–97.

Schierenbeck, K. A., M. Skupski, D. Lieberman & M. Lieberman, 1997. Population structure and genetic diversity in four tropical tree species in Costa Rica. Molec. Ecol. 6: 137–144.

Tingey, S. V. & J. P. Del Tufo, 1993. genetic analysis with random amplified polymorphic DNA markers. Plant Physiol. 101: 349–352.

Tomlinson, P. B., 1986. The botany of mangroves. Cambridge University press, Cambridge.

Triest, L., 1991. Conservation of genetic diversity in water plants. Op. Bot. Belg. 4: 241–258.

Welsh, J. & M. McClelland, 1990. Fingerprinting genomes using PCR with arbitrary primers. Nucleic Acids Res. 18: 7213–7218.

Welsh, J., R. J. Honeycutt, M. McClelland & B. W. S. Sobral, 1991. Parentage determination in maize hybrids using the arbitrarily primed polymerase chain reaction (AP-PCR). Theor. Appl. Genet. 82: 473–476.

Williams, J. G. K. A. R., K. J. Kubelik, J. A. Livak, S. Tingey & V. Rafalski, 1990. DNA polymorphism amplified by arbitrary primers are useful as genetic markers. Nucleic Acids Res. 18: 6531–6535.

Hydrobiologia **413:** 17–33, 1999.
R.S. Dodd (ed.), Diversity and Function in Mangrove Ecosystems.

Faunal zonation and assemblages in the Pacific Colombian mangroves

Jaime R. Cantera K.[1], Bernard A. Thomassin[2,*] & Patrick M. Arnaud[2]
[1]*Departamento de Biologia, Universidad del Valle, Apartado aereo 25360, Cali, Colombia*
E-mail: jcantera@biomarina.univalle.edu.co
[2]*Centre d'Océanologie de Marseille, Station marine d'Endoume (Unité CNRS/DIMAR), Université de la Méditerranée, traverse de la Batterie des Lions, 13007 Marseille, France*
E-mail: thomassi@com.univ-mrs.fr
*(*Author for correspondence)*

Key words: benthos zonation, mangroves, East Pacific, Colombia

Abstract

The zonations of trees and marine associated fauna of two mangrove areas of the Pacific coast of Colombia, Malaga and Buenaventura bays, are compared. These bays are exposed to huge terrigeneous inputs, abundant rainfall, high tidal range and, for the later, to deforestation, urban and industrial pollution. Vertical distribution of the macrofaunal assemblages is under the influences of wetness and tidal gradients. These assemblages are associated with the various substrates (mangal substrates: canopy, trunks and roots; associated soil substrates: more or less clogged and muddy sands; litter) and with the tidal features. As their specific composition is more dependent on the ethology of each species (air-exposure, feeding, reproduction...), the zonation appears more quantitative than qualitative, the same species occurring frequently in different zones or subzones. Basically, similar assemblages are observed in the two bays. They can be locally enriched by species restricted to particular edaphic conditions such as gravelly muddy flats and rocks at Malaga, and intertidal creeks and running waters at Buenaventura, where they are relatively less influenced by moderate human impacts.

Introduction

Tropical mangroves biotopes are mainly located in the intertidal area which can be quite large along the shores due to local tidal range and topography. Mangal trees, as a result of their height, represent an ecotone between the terrestrial (high canopy) and the marine environment (lower parts of trunks and roots). Their bases occupy mainly two zones, originally defined along the rocky shores (Pérès, 1982): the supralittoral and midlittoral zones. The supralittoral zone is washed by seawater only at high tide or by splashing. The midlittoral zone extends in the central intertidal area and it can be divided into upper, middle and lower subzones. Landwards, the spray and the atmospheric moisture enable the colonization of a third zone: the adlittoral, influenced by the terrestrial ecosystems. Seawards, the lower midlittoral fringe is sometimes occupied by stoloniferous roots and small pioneer mangal trees.

Each mangal biotope can be located in more than one zone and/or subzone depending on the bottom slope and height of trees. For example, on the Pacific coast of Colombia, the leaves and stems can be located in the adlittoral, supralittoral and upper midlittoral zones, depending of the location and height of the trees.

The purpose of this paper is to point out the existence of a zonation of the marine macrofauna, as well as the occurrence of species characteristic of the mangal habitats of the Colombian Pacific coast.

Along the South American Pacific coast, mangrove swamps grow between about 7° N (Jurado, in Colombia: Prahl et al., 1990) and 3° 35′ S (rio Piura, in Peru: Otazu Abrill et al., 1995). In Colombia, they are the predominant coastal ecosystem, the mangroves extending up to 2837 km^2 and lining 850 km out of 1300 km of shoreline (Cantera, 1991) (Figure 1A). This results from: (a), the huge terrigeneous sedimentation (clays and muddy sands) on a continental shelf enlarging towards the South with many estuaries (Mo-

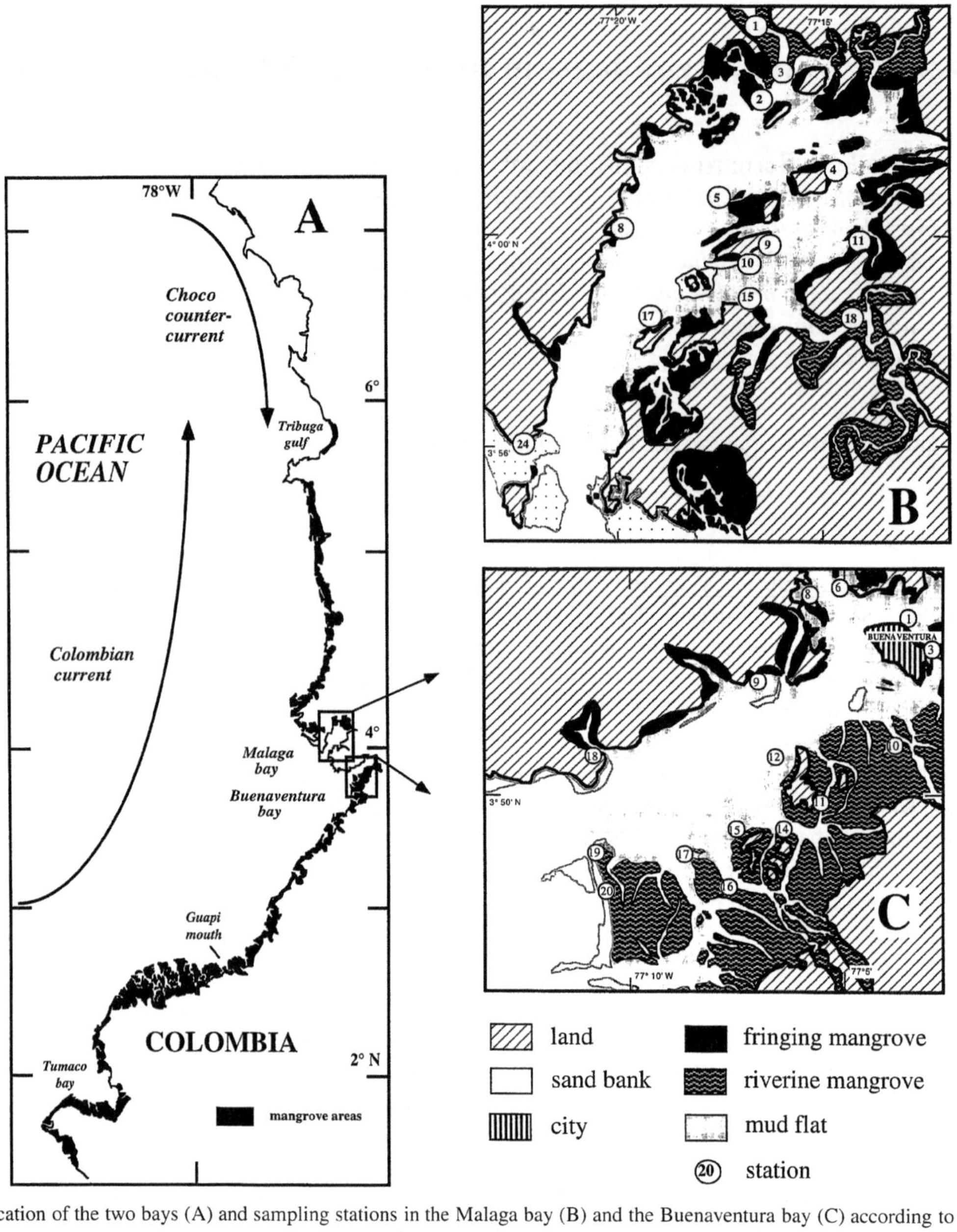

Figure 1. Location of the two bays (A) and sampling stations in the Malaga bay (B) and the Buenaventura bay (C) according to the mangrove types.

lina & Mirmand, 1992; Pujos et al., 1995); (b), the wet climate characterized by abundant and frequent rainfalls (3–5 m yr^{-1}), without a dry season, associated with a relatively high and constant air temperature (25.1 °C±2 °C), and to constant but weak winds; (c), high water temperatures (25–26 °C), lowered salinities (20–33.5 psu) and high turbidity, associated with coastal currents distributing the particles (current of Colombia and counter-current of Choco in the northern area) and (d), high tidal range (up to 3.7 m at Buenaventura, up to 4 m at Malaga). The extension of these Colombian Pacific mangroves decreases as the result of human activities (exploitation and pollution) (Alvarez-Leon, 1990).

Two representative sheltered bays of the central Colombian coast were selected for this study, according to (a), the diversity and the development of their mangroves, (b), the good knowledge of their fauna, and (c), their economic importance: the Malaga bay and the Buenaventura bay.

These bays are located in the biogeographic Province of Panama, that extends between the Fonseca Gulf in Salvador (13° 16′ N, 87° 30′ W) and the mouth of the Tumbes river, in Peru (3° 38′ N, 80° 20′ W). The mangroves growing in Colombia belong to the W. American-African floristic complex defined by Chapman (1977), but some endemic mangals are also present (West, 1956, 1977; Prahl et al., 1990; Cantera, 1991). They include up to nine trees with four other true mangrove associated arborescent shrubs or ferns and associated terrestrial plants (Cantera & Arnaud, 1997).

Physiographic features of the selected mangroves

The Malaga bay

The Malaga bay (4° N, 77° 20′ W), 126 km^2, with an average depth of 12–15 m, and under low human impacts at the time of this study (1985–1988) (Figure 1B). Mangroves predominate in the eastern areas, mainly at the mouths of the rivers, their extension along the western shores being limited by the Tertiary cliffs. Riverine (ripicolous) mangroves are favoured by terrigeneous alluvial deposits developed by large rivers. The mangal trees, up to 40 m high, are dominated by *Rhizophora harrisonii* and *R. mangle*, with some *Avicennia germinans*, and towards the inner parts, *Pelliciera rhizophorae*, eventually mixed with *Mora megistosperma*. The fringing mangroves of the SE and E of the bay are dominated on the outer border by *R. mangle*, in the central areas by mixed *R. mangle*, *A. germinans* and *Laguncularia racemosa*, and in the inner areas by *M. megistosperma* and *P. rhizophorae*. Wide intertidal bare muddy flats occur seawards. In the NW of the bay and around the central islands, patchy fringing mangroves dominated by *P. rhizophorae* are observed. The more muddy substrates are colonized by *R. harrisonii* (10–15 m high) in the inner areas and by smaller *R. mangle* (less than 5 m high) in the outer areas. At the foot of the high cliffs at the entrance of the bay, patches of dwarf fringing mangroves grow.

The Buenaventura bay

The Buenaventura bay (3° 54′ N, 77° 5′ W), located 36 km South, a little smaller, 96 km^2, with an average depth of 15 m, and submitted to deforestation and to urban and industrial perturbation and pollution of the main Colombian harbour of this coast (Vegas Vélez et al., 1976) (Figure 1C). The mangroves are also distributed all around the bay, but they are better developed in its southern areas. There, large riverine mangroves are dominated by *R. harrisonii* (up to 30 m high) and *A. germinans*, and in the inner areas some *P. rhizophorae* and *M. megistosperma* are scattered. At the entrance of the bay, the mangrove of Punta Soldado is growing behind an eolian sand spit, with a mixture of *R. mangle*, *R. harrisonii*, *A. germinans* and *L. racemosa* seawards, and *Conocarpus erectus* and *Hibiscus tiliaceus* landwards. Along the northern shores of the bay, mangroves look like the fringing mangroves of Malaga bay. These small mangroves, with *R. mangle* and *A. germinans*, are growing around Cangrejo Island and locally around the Tertiary hard substrates.

In the mangroves of both bays, under the influence of river outflows and rainfalls, the surface water salinity varies between 19 and 8 psu (except at the mouth of the large Malaga bay), without a clear inner/outer gradient (Table 1). Average water surface temperature reaches 27.8–27.4 °C.

Typical distribution of the mangal trees along transects through riverine and dwarf fringing mangroves is sketched in Figure 2. These mangroves are a patchwork of soft muddy bottoms and hard substrates or similar: roots, trunks and canopy of the mangal trees (associated or not with rocky outcrops, detrital accumulations, or litter and wood deposits), and dried bare muddy bottoms. The mangrove fauna is, therefore, a mixture of species living: (a) on (epibiotic) or inside (burrowing and boring) the hard sustrates, (b) on or in the soft substrates, and (c), on or below the litter.

Material and methods

Samples

The study was conducted between June 1985 and December 1988, at 17 stations in Malaga bay and 16 at Buenaventura bay (Figure 1). The location, mangrove types, water salinities and sedimentological parameters of the stations are given in Table 1.

Samples were collected from the lower midlittoral subzone up to the adlittoral zone, along transects perpendicular to the shore, at every 10 m, on roots, trunks and canopy of the mangals. The mangrove terrestrial fauna was not taken into account. Each station was sampled one time, so no eventual seasonal variation was taken in account.

Table 1. Describing parameters of the stations in the Malaga (M..) and Buenaventura (B..) bays. Mangrove types: DR, dwarf fringing on rocks; FR, fringing; RP, riverine. Nature of perturbations: D, deforestation; HD, heavy deforestation; P, pollution. Wave exposure: along a scale from 1, sheltered, to 3, more exposed

Location			Mangrove		Water		Sediment components (%) according to zonation																			
							Supralittoral					Upper Midlittoral					Middle Midlittoral					Lower Midlittoral				
Station number	Name	Distance from open sea (km)	Type	Perturbation	Exposure	Salinity (ppt)	Gravels+sands	Fine sands	Silts+clays	Water content	Organic matter	Gravels+sands	Fine sands	Silts+clays	Water content	Organic matter	Gravels+sands	Fine sands	Silts+clays	Water content	Organic matter	Gravels+sands	Fine sands	Silts+clays	Water content	Organic matter
M24C	Juanchaco	2.4	FR		2	23.2						52	45	3	19	2	47	7	28	2						
M17	Isla Monos	9.4	FR	D	3																					
M8A	Pta. La Muerte	11.2	FR		1	23.0	72	19	11	11	3	61	25	15	18	3	57	22	20	22	3					
M8B	Pta. La Muerte	11.4	FR		1	21.0																				
M15	Bahia Caracas	15.6	FR		3	16.8						78	16	6	14	2										
M5	N I. Grande	16.2	DR		2	13.7						30	70	0	15	2										
M10	Isla Cabezon	16.8	DR	D	2																					
M9	N Isla Cabezor	18.0	FR	D	2																					
M11	Cabo Viatana	18.6	FR	P	2																					
M4B	A. La Plata	19.8	FR	HD	1	11.8											34	56	10	20	2					
M4A	A. La Plata	20.0	FR		1	12.7																				
M4C	A. La Plata	20.6	FR		1	15.8																				
M2	E. Corozal	22.1	RP		1	13.4	37	54	10	13	3	29	62	9	15	4	33	61	7	24	3	29	60	12	33	3
M18	Bahia Sierpe	22.2	FR	D	1																					
M3B	Isla Aguante	22.4	DR		1	16.8																				
M3A	Isla Aguante	22.6	DR		1	14.5	32	57	11	19	2	31	48	40	24	2	28	62	11	28	2	26	63	11	32	3
M1	Q. Valencia	26.1	RP		1	10.3	43	45	13	10	2	36	53	11	14	3	30	59	7	22	2					

Table 1. Continued

Station number	Name	Distance from open sea (km)	Type	Perturbation	Exposure	Salinity (ppt)	Gravels+sands	Fine sands	Silts+clays	Water content	Organic matter	Gravels+sands	Fine sands	Silts+clays	Water content	Organic matter	Gravels+sands	Fine sands	Silts+clays	Water content	Organic matter	Gravels+sands	Fine sands	Silts+clays	Water content	Organic matter
Location			Mangrove		Water		Sediment components (%) according to zonation																			
							Supralittoral					Upper Midlittoral					Middle Midlittoral					Lower Midlittoral				
B19B	Pta. Soldado	2.6	FR	HD	2	16.2	77	16	7	14	4	73	17	10	20	4	69	17	13	37	5					
B20	E. Veneno	3.0	RP	D	1	17.2											15	67	17	33	4	22	63	16	40	4
B18	Pta. Bazan	3.9	FR	HD	3	13.2											83	10	7	32	1	90	8	3	39	1
B17	Rio Anchicaya	6.6	RP	D	2	11.6	37	49	14	17	1	33	53	14	22	3	35	53	13	33	2	42	50	9	41	2
B16	Rio Anchicaya	6.6	RP	D	1	8.6																38	51	10	40	2
B15	Rio Potedo	8.4	RP	D	2	11.2	33	53	14	18	2	45	42	13	20	2	37	48	15	34	2					
B14	Rio Potedo	8.2	RP	D	1	14.3											48	39	14	31		45	37	18	41	1
B12	Pta. Limones	12.9	RP	HD	2	14.3											31	15	5	31	3	29	55	17	40	2
B9A	Pta. Arenas	12.4	FR?/HD	2	18.9						83	7	10	14	3	88	6	7	28	3						
B11	Rio Limones	12.6	RP	HD	1	12.1											44	44	13	33	3	41	51	8	40	2
B8	E. Pichiron B	16.1	FR	P/D	1	16.7																				
B10	Rio Dagua	17.4	RP	HD	1	8.2											53	34	14	32	2	44	44	13	39	
B6	E. Pichiron A	18.8	FR	P/D	1	18.6						35	49	16	14	1	37	49	15	20	1					
B3A	E. San Antonic	19.0	RP?/HD	1	10.7	25	27	47	13	1	22	55	23	16	2											
B3B	E. San Antonic	19.3	RP?/HD	1	10.2																					
B1	Pta. Muneton	19.2	FR?/HD	1	14.6	29	20	53	10	2	18	60	22	13	2	28	58	14	18	1						

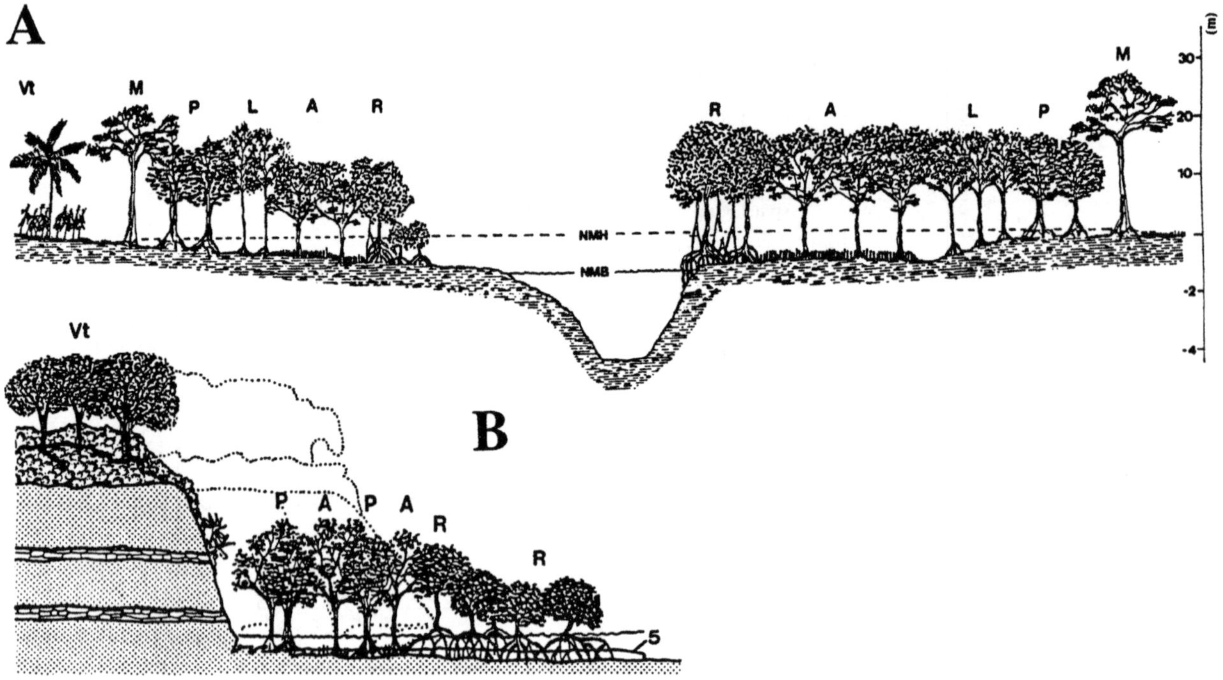

Figure 2. Sketches of mangal-trees zonation in a ripicolous mangrove (Anchicaya river, Buenaventura bay) (A), and in a fringing mangrove (Malaga bay) (B). A: *Avicennia*, L: *Laguncularia*, M: *Mora*, P: *Pelliciera*, R: *Rhizophora*, Vt: terrestrial vegetation; NHB: mean low tide level; NMH: mean high tide level.

For the canopy, trunks and roots, two circular areas (1.1 m in diameter), one horizontal and one vertical, were surveyed on the transect line and all animals living inside these two planes collected. Densities of the highly mobile crabs were estimated through three complementary methods: (a), counting in the circles, (b), counting from a distance, using binoculars, in a delimited surface and (c), using marking capture and recapture methods.

For muds, quantitative samples were taken by shoveling from 25×25 cm surfaces down to 20 cm deep, and then fixing the sediments on the field in 10% buffered formaldehyde. Sediment macrofauna was extracted by sieving through 2 mm and 1 mm square-mesh.

Data treatment

Species from all samples were identified and counted (or, for colonial organisms, measured for estimation of their surface coverage). Identifications were later confirmed by taxonomists (see 'Acknowledgments'). For each type of substrates (canopy, roots and trunks, and muds) data were pooled in matrices by species and samples and then analyzed using cluster and multivariate analyses (principal component analysis, or PCA, and correspondance factorial analysis, or CFA; Benzecri, 1973) to check for an eventual continuum between the different zones. Then qualitative and quantitative comparisons between stations or tidal levels were made using classical biological indices: presence (p), species richness (S), abundance (A), frequency (f%), Shannon's diversity index (H′) and Pielou's equitability (J′). Similarity between stations and fauna was evaluated through Jaccard's index.

All these methods and indices are classical (Thomassin, 1978a; Snedaker & Snedaker, 1984). The methodological originality of the present work is to combine them in order to point out more accurate and comprehensive results.

Recognition of a zonation in the mangrove assemblages

Depending on their height and form, the canopies, trunks and roots of the mangal trees are exposed temporarily to various wetness conditions, either aerial or marine. In the two bays, animals were classified according to the three main substrates of the mangroves: 1. canopy, 2. trunks and roots and 3. muddy substrates. To check for the existence of a continuum or

Table 2. Contribution to the variance of the first three axis in the FCA analyses

	Axis 1 (%)	Axis 2 (%)	Axis 3 (%)
Branches	55.6	16.0	8.4
Roots	25.1	17.4	10.9
Soils	27.2	10.9	6.7

of a zonation, the faunistic data were analyzed separately according to these substrates using multivariate analysis and clustering.

CFA analyses

Since they gave results similar to those of the PCA analyses, CFA analyses alone are presented here. From them, it is observed that the contribution of the first 3 axis to the total variance varies substantially with the substrates (Table 2).

For canopy (branches), the axis 1 represents more than 55%, far more than the contribution of the other axis, pointing out a strong gradient (Figure 3). Most of the samples located in the supralittoral zone are aggregated and opposite to those located in the middle midlittoral subzone. The barycenter of the plane 1–2 is occupied by those of the upper midlittoral zone. Even if the general distribution follows a Guttman's effect, these groups of samples are sufficiently distinct to disprove the presence of a continuum through the zones. Distribution along the axis 3 (8%) separates a group of stations located in the disturbed mangroves of both Malaga and Buenaventura bays (sta. M4B, M4C, B1, B3A, B3B) from the other ones.

For roots and muddy soils, the contribution of axis 1 is less remarkable (25 and 27%, respectively), but contributions of axis F2 and F3 are not increased accordingly (Figures 4 & 5). For both substrates, samples belonging to each zone and/or subzone are clearly gathered in clouds demonstrating the existence of zonation. Furthermore, in the muds, distribution of stations along the axis F2 in plan 1–2 is linked with the presence or absence of gravels or coarse sands.

In conclusion, all these CFA analyses demonstrate that there is indeed a zonation according to tidal levels and wetness. Consequently this zonation will be used throughout the following data treatment.

Cluster analyses

Clustering of the samples according to the three substrates and various zones shows that the similarity index, based on presence-absence or/and abundance, is homogeneous and reaches very high values (up to 80–95%).

From the supralittoral zone fauna of the trees (canopy, roots and lower parts of trunks), dendrograms distinguish the stations according to the two bays and to five types of mangroves (Figure 6A). They are: 1. dwarf mangroves on rocks, 2. fringing mangroves on gravelly substrate, 3. mangroves exposed to deforestation, 4. fringing mangroves periodically invaded by sands, and 5. typical riverine mangroves. But from the midlittoral zone fauna of the trees (roots), clusters of the data of the two bays and of the various types of mangroves are more mixed (Figure 6B). From the mud fauna, clusters distinguish better the two bays and secondarily take into account some sedimentological features (presence of gravels, sands or litter).

As a result, the different types of mangroves will not be separated in the analysis of the faunistic assemblages of the three substrates and of the zones and subzones.

Faunistic assemblages

The tree fauna (Table 3, Figure 6)

The canopy assemblage

From the canopy, only 14 species were recorded, the commonest were the gastropods *Littorina aberrans*, *L. zebra*, *L. fasciata* and *L. varia*, the crabs *Aratus pisonii*, *Pachygrapsus transversus* and *Goniopsis pulchra*. Several of them occur also on the rocky cliffs of the two bays.

In the highest canopy of the *Rhizophora* and *Avicennia* belt, the most abundant species is *A. pisonii*, always present (mean density=5.7 ind m^{-2}; maximum frequency f_{max}=60%). In the upper branches of the trees and ferns growing landwards (*Laguncularia racemosa* and *Acrostrichium aureum*), *L. aberrans* occurs in most stations (mean density=1.6 ind m^{-2}; f_{max}=22%). The crabs *Sesarma occidentale* and *S. rhizophorae*, that climb during the rains to feed on the debris of leaves accumulated in the epiphytic Bromeliads, are uncommon (f<6%) and were never observed on roots. On the contrary, *Sesarma aequatoriale* ($f \leq$8%) also inhabits on the roots. These crabs

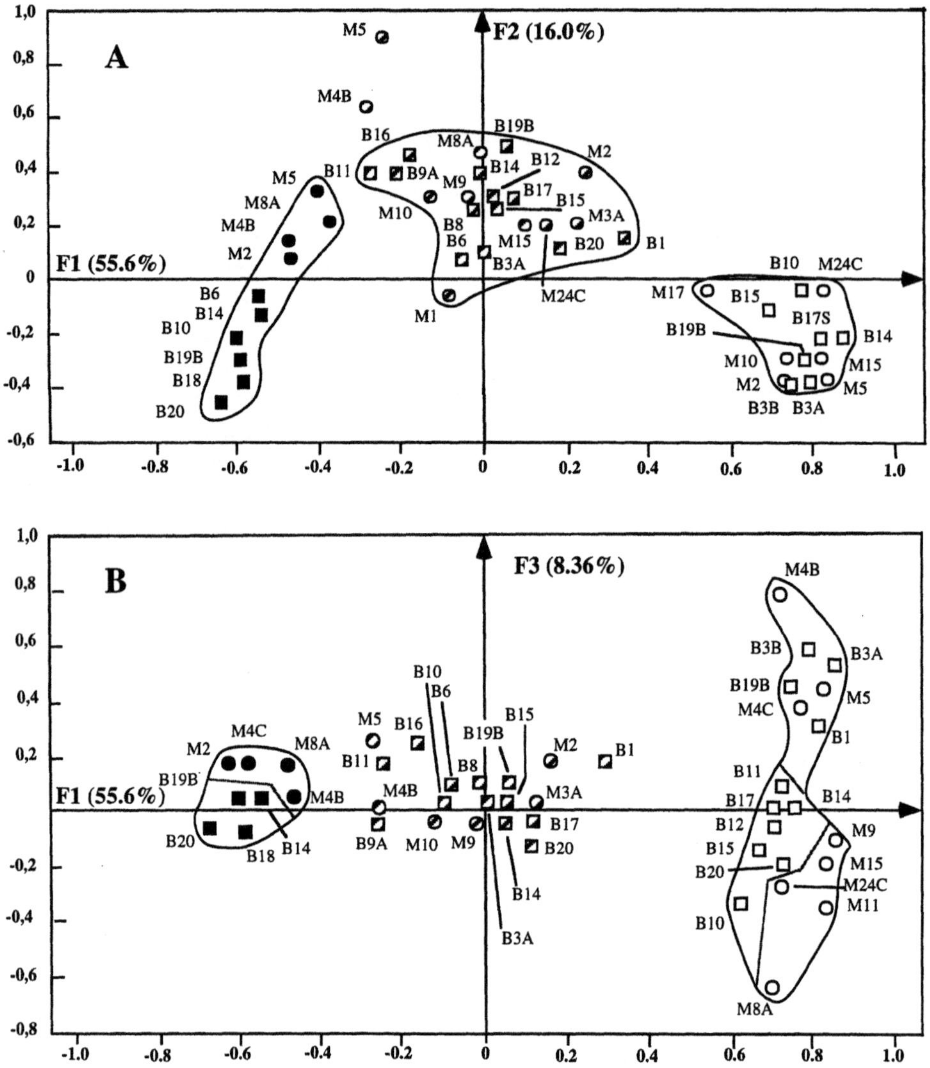

Figure 3. Correspondance factorial analysis (CFA) graphs of the canopy samples according to the bays and zones. (A) Distribution of the stations in the plane F1–F2; (B) Same in the plane F1–F3.

are not exclusive of mangroves and also occur on coastal works or in the coastal terrestrial environment. In samples, the diversity index (H′) varies from 0.35 to 1.7 bit ind.$^{-1}$ (H′ canopy=1.65 bit ind.$^{-1}$) and the equitability index (J′) from 0.2 to 0.9 (J′ canopy=0.59); lower values being recorded in dwarf mangroves of the Malaga bay (La Plata, sta. M3A, B), where *L. aberrans* and *A. pisonii* predominate, while higher values are observed in riverine mangroves, particularly in the Buenaventura bay (Pta. Soldado, sta. B19B).

When the canopy is located in the high tidal zone (supralittoral zone), gastropods predominate in number (*f*=66%), but crab species are more numerous (7 spp.). Diversity is higher in Buenaventura bay than in Malaga bay (H′ branches=1.73 against 1.43 bit ind.$^{-1}$; J′ branches=0.58 against 0.43). In both bays, the most abundant and frequent species are *L. zebra* (f_{max}=65) and *A. pisonii* (f_{max}=36). From the mouth of bays landwards, the species richness increases while the faunistic assemblages become less distinct (H′=0.9–1.9 bit ind.$^{-1}$; J′=0.4–0.8).

When the branches are located in the upper intertidal zone (upper and middle midlittoral subzones), gastropods (*f*=57–40%) and brachyurans (*f*=44–60%) predominate also. Species richness is close to that of the highest part of canopy (adlittoral and supralittoral zones), but abundances (A′) are greater due to an increase in wetness. So the diversity index is higher than

Table 3. Distribution of the faunistic assemblages on branches and roots according to zonation in the Buenaventura (BUEN.) and Malaga (MAL.) bays. Adlitt.: adlittoral zone; Lower Mid.: lower midlittoral subzone; p: observed presence; p′: relative presence calculated for 9 stations; f%: frequency

Zone or Subzone	Adlitt.		Supralittoral								Upper midlittoral								Middle midlittoral								Lower Mid.	
Biotopes	Branches		Branches				Roots				Branches				Roots				Branches				Roots				Roots	
Location	BUEN.		MAL.		BUEN.		MAL.		BUEN.		MAL.		BUEN.		MAL.		BUEN.		MAL.		BUEN.		MAL.		BUEN.		BUEN.	
Species	p′	f%	p′	f%	p	f%	p	f%	p	f%	p	f%	p	f%	p	f%	p	f%	p	f%	p	f%	p	f%	p	f%	p	f%
Sesarma occidentale	4	5.7			2	0.4																						
Sesarma rhizophorae			2	0.2																								
Sesarma angustum			1	<0.1																								
Littorina aberrans	7	21.9	6	2.3	9	4.7	2	0.4	2	0.6	2	0.2																
Aratus pisonii	9	60.1	9	25.0	9	36.6	9	56.9	9	46.0	7	7.8		9.4	9	4.5	9	4.5	5	0.1	2	0.1	3	<0.1	3	<0.1		
Littorina zebra	7	4	8	65.5	9	49.4	5	3.0	6	3.5	9	21.5	9	20.3	9	43.0	9	42.0	2	0.2	1	0.1	7	3.8	9	2.7	2	<0.1
Sesarma aequatoriale	2	8.2	3	1.8	5	1.6	3	4.9	1	0.7	3	0.3	2	0.5	3	2.0												
Pachygrapsus transversus	2	0.1	7	3.5	9	4.1	9	23.9	9	34.4	9	21.1	9	27.9	9	17.6	9	20.3	9	44.0	9	30.0	9	21.1	9	18.0	9	22.4
Goniopsis pulchra	1	<0.1	4	0.1	5	0.2	9	10.1	8	8.7	8	11.5	9	10.3	9	8.9	9	7.9	7	16.1	9	29.0	9	14.1	9	17.9	9	18.6
Littorina fasciata			4	1.6	9	3.1	2	0.8	5	6.1	9	29.6	9	26.2	9	15.2	9	18.8	9	11.5	9	5.8	9	4.7	9	6.5	4	0.1
Chthamalus panamensis			2	0.1							5	0.8	2	0.5					3	0.2	2	0.2						
Littorina varia									1	<0.1	9	5.3	9	4.2	5	0.3	4	0.3	9	27.3	9	34.7	9	13.7	9	18.8	9	13.4
Teredinidae sp. 1											5	1.7	2	0.1	4	0.4	3	0.2	5	0.6	1	<0.1	2	0.1	1	<0.1		
Tetraclita panamensis											2	0.3	2	0.5					1	0.1								
Crassostrea corteziensis															7	5.9	9	5.4					9	8.3	9	3.9	9	0.4
Thais kiosquiformis															4	0.2	6	0.1					9	23.4	9	22.5	9	19.8
Nausitora dias															1	<0.1	1	0.1									2	<0.1
Martesia striata															3	0.1	1	<0.1					5	0.2	1	<0.1	1	<0.1
Bankia gouldi															2	0.1							4	0.2				
Eurypanopeus planus															4	0.2	3	0.1					3	0.2	3	0.1		
Chthamalus sp. 1															2	0.3	1	0.3					2	0.1	1	0.1		
Neanthes galetae															5	1.4							4	0.2				
Nerita scabricosta															2	<0.1							2	0.6				
Teredo furcigera																	1	<0.1					1	<0.1	2	0.1	2	0.1
Clibanarius panamensis																							9	3.9	9	5.3	9	5.8
Talitrus sp. 1																							9	1.8	8	1.9	9	5.4
Cerithidea mazatlanica																							6	1.1	6	0.2	9	1.5
Anachis nigricans																							5	0.5	3	0.1	3	0.2
Cerithidea pulchra																							2	0.2	3	0.2	2	0.1
Syllidae sp. 1																							4	0.1				
Mystides sp. 1																							4	<0.1				
Thais biserialis																							3	<0.1				
Petrolisthes armatus																							3	0.1	1	0.1		
Pododesmus foliatus																							2	1.2	1	0.5		
Nereis riisei																							2	0.1	1	0.1		
Lysidice nineta																							2	0.4	1	<0.1		
Crepidula arenata																							1	<0.1	1	0.1		
Theodoxus luteofasciatus																									1	0.2		
Palaemonetes hiltoni																									1	0.1		
Uca vocator																									2	0.7	1	<0.1
number of samples	11		13		12		9		10		11		14		12		14		9		13		9		13		9	
total species richness S	7		10		8		7		8		11		10		17		14		9		9		29		26		16	
Total abundance A		6600		21636		23776		4476		4962		23278		28298		39011		44552		25399		41223		58379		85992		70626
relative total abundance A′ (9		5400		14979		17832		4476		4466		19046		18192		29258		28641		25399		28539		58379		59533		70626
biotope diversity index H′	1.65		1.43		1.73		1.74		1.86		2.56		2.46		2.46		2.26		1.92		1.84		3.11		2.95		2.41	
biotope equitability index J′	0.59		0.43		0.58		0.62		0.62		0.74		0.74		0.60		0.59		0.60		0.58		0.64		0.63		0.60	

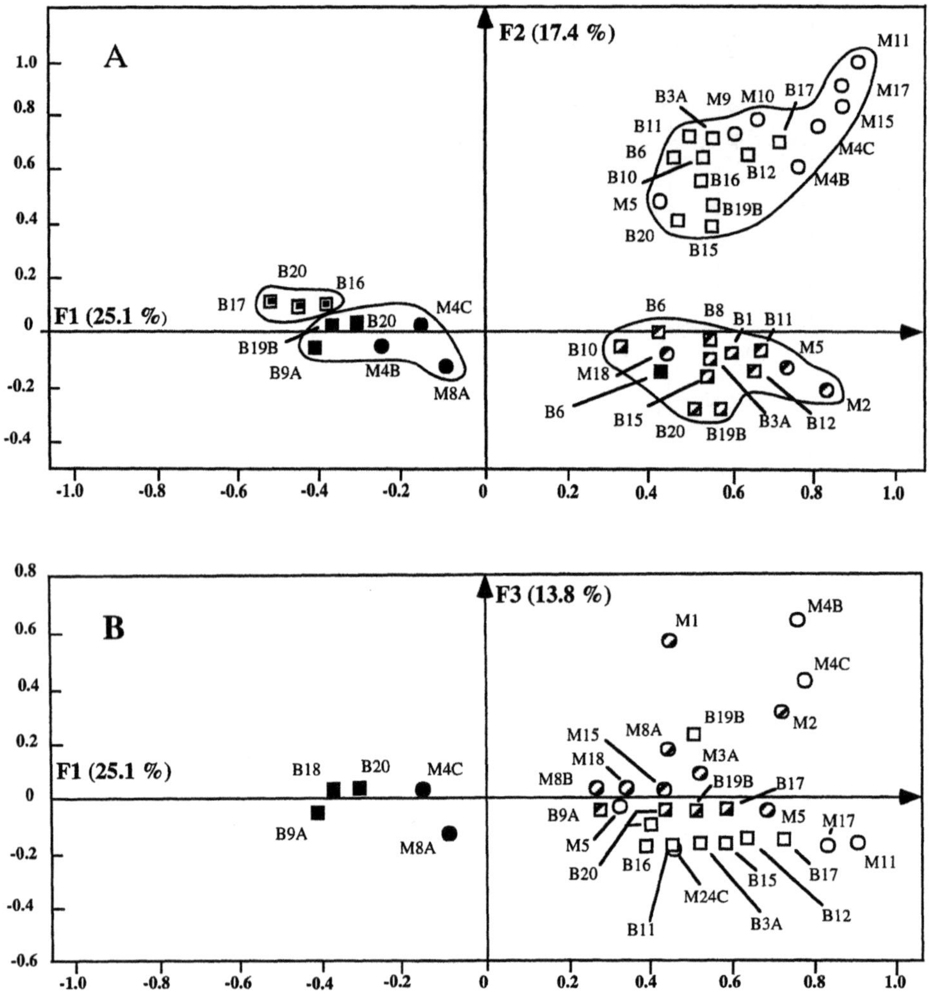

Figure 4. Correspondance factorial analysis (CFA) graphs of the root samples according to the bays and zones. (A) Distribution of the stations in the plane F1–F2; (B) Same in the plane F1–F3. Signs used as in Figure 3.

in the supralittoral zone (H′=1.8–2.5 bit ind.$^{-1}$; H′ branches=2.6–1.9 bit ind.$^{-1}$). In both bays, the most common species of the upper midlittoral subzone are *L. fasciata*, *L. zebra* and *P. transversus*, while in the middle midlittoral subzone *L. varia* takes the place of *L. zebra*.

The trunk and root assemblage

In this biotope, 35 species were recorded, molluscs dominate in species richness (11 spp. of gastropods and 7 of bivalves), while crabs and gastropods are the most abundant.

In the intertidal zone, roots and lower parts of the trunks are exposed to a wetness gradient depending on the exposure to winds and tides, and also on their algal-turf coverage. Therefore, the zonation of the mangal trees plays an important role in the macrofaunal distribution.

Roots are commonly intertidal (upper, mid and lower midlittoral zones), but the higher aerial roots of *Rhizophora* and the roots of *Mora* and *Pelliciera* are located in the supralittoral zone. In this last zone, 8 species are recorded, the crabs *A. pisonii* and *P. transversus* being dominant, and the abundance is 3–4 times lower than in the canopy. On the aerial roots of *Rhizophora* submerged only at high tide, *L. zebra* is dominant (*f*=42%) followed by *P. transversus* and *L. fasciata*. Oysters (*Crassostrea corteziensis*) and various wood-boring bivalves (Teredinidae) are generally present but with low frequencies. Diversity and equitability indices are similar to those observed in the upper midlittoral subzone of canopy.

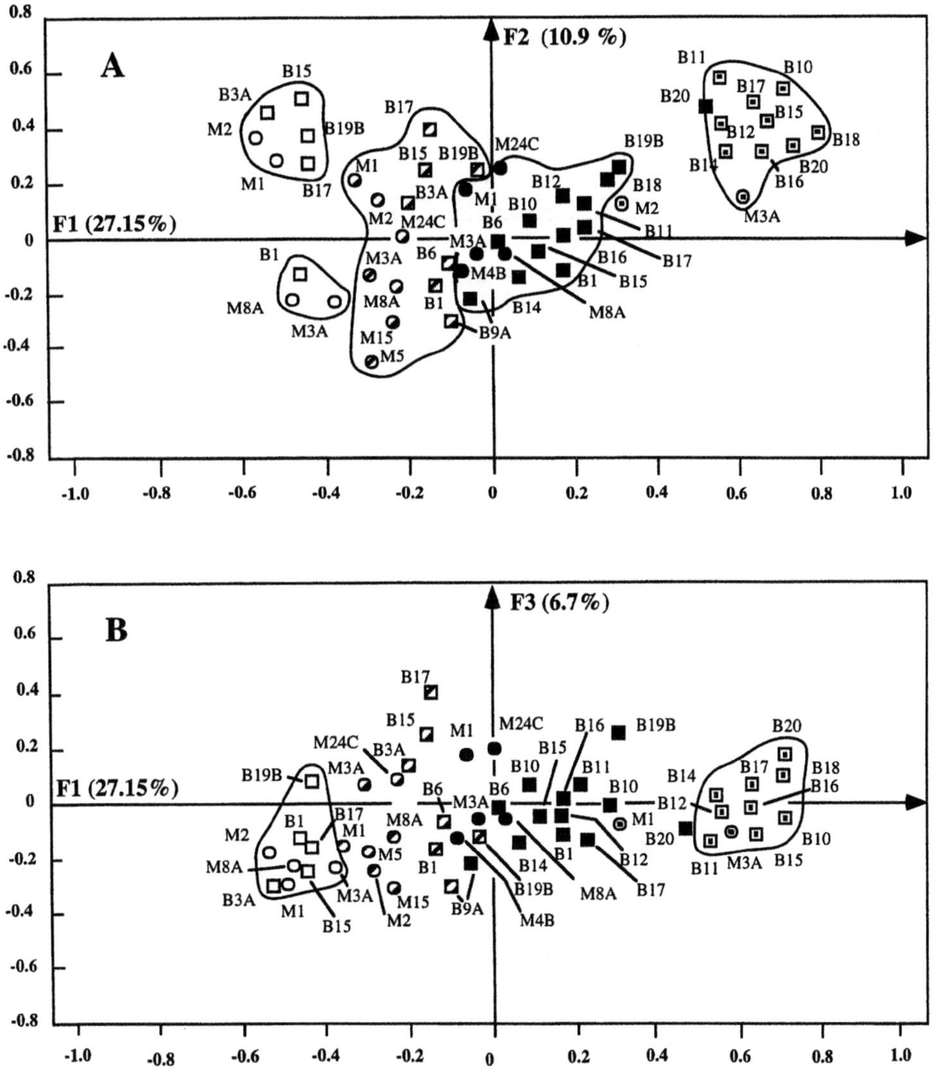

Figure 5. Correspondance factorial analysis (CFA) graphs of the soil samples according to the bays and zones. (A) Distribution of the stations in the plane F1–F2; (B) Same in the plane F1–F3. Signs used as in Figure 3.

On the roots located in the middle midlittoral subzone, where moisture is increased by algal-turfs and canopy shade, species richness increases to 32 species (11 gastropods, 6 bivalves, 5 polychaetes and 10 crustaceans). Gastropods are the most abundant (50%), followed by crabs (36%). Dominant species are the same in the two bays: *Thais kiosquiformis*, *L. varia*, *P. transversus* and *G. pulchra*. As on roots of the upper midlittoral subzone, oysters and wood-boring bivalves are encountered. The increase in wetness in the cork of roots and in the algal-turfs trapping muddy sediments, and the presence of more sheltered habitats enable several other species to live there: *Clibanarius panamensis*, *Talitrus* sp. 1, *Cerithidea mazatlanica* and others. The diversity index is far higher (H′ biotope close to 3 bit ind.$^{-1}$) than in the surrounding zones. As in the other zones, its lowest values are reached in the stations close to the Buenaventura city, whereas its highest are recorded at the mouth of the rivers.

In some eroded edges of mangrove creeks, roots emerge from the sediment in the lower midlittoral subzone. In Buenaventura bay, only 16 species live on these roots, the dominant ones being similar to those of the middle midlittoral subzone. Lowering of the species richness may be explained tentatively by an increase in the mud deposition and clogging.

Table 4. Distribution of the faunistic assemblages in the muddy substrates according to zonation in the Buenaventura (BUEN.) and Malaga bays. p: observed presence; p′: relative presence calculated for 5 stations; f%: frequency

Zone or Subzone	Supralittoral				Upper midlittoral				Middle midlittoral				Lower midlittoral	
location	MALAGA		BUEN.		MALAGA		BUEN.		MALAGA		BUEN.		BUEN.	
species	p′	f%	p	f%	p′	f%	p′	f%	p′	f%	p′	f%	p′	f%
Sesarma aequatoriale	5	28.6	4	31.7	3	0.7	3	1.1						
Marinula sp. 1	1	37.8	2	10.6			2	0.3			2	<0.1		
Cerithium stercusmuscarum	4	6.3			3	6.8			3	0.5				
Uca zacae	1	2.5			1	0.3								
Uca panamensis	1	1.7			1	0.3								
Modulus catenulatus	1	0.7			1	5.0								
Pedipes sp. 1			1	11.0			1	0.7						
Uca stylifera			1	7.2			1	0.7						
Uca heteropleura			1	1.0	1	<0.1	1	0.1						
Panopeus purpureus	5	15.3	5	12.4	5	16.2	5	13.8	5	18.8	5	11.8	5	10.8
Pachygrapsus transversus	5	4.9	5	7.7	4	7.2	5	6.0	3	6.5	5	6.2	5	7.0
Clibanarius panamensis	3	0.8	2	2.0	5	14.2	5	19.5	5	14.3	5	16.4	5	14.0
Littorina varia			1	0.5	4	9.0	4	1.2	3	0.4	2	<0.1	1	<0.1
Natica unifasciata			1	0.5	2	0.2	1	0.6	2	0.2	2	0.2	2	0.2
Talitrus sp. 1	1	1.5	1	5.4	1	12.9	1	35.0	3	6.3	1	5.3		
Polymesoda inflata			1	5.5			3	1.2			2	0.2		
Melampus tabogensis			1	4.6			1	0.5			0	<0.1		
Corbula biradiata					1	0.3								
Cerithidea pulchra					1	1.0	1	0.7	2	2.0	0	0.2		
Poecilochaetus sp. 1					2	0.2	1	<0.1	3	0.6	0	<0.1		
Anachis rugosa					1	0.3	1	<0.1	3	2.1				
Cerithidea valida					1	0.3	1	<0.1	1	<0.1				
Rhinocoryne humboldti					1	1.6			2	0.3				
Anomia fidenas					1	2.5			1	<0.1				
Corbula bicarinata					1	<0.1			1	<0.1				
Ucides occidentalis					1	0.2			1	0.4				
Leptalpheus mexicanus					1	0.2			1	0.2				
Leucosilla jirunei					1	<0.1			1	<0.1				
Goniopsis pulchra					5	10.0	4	4.9	5	15.3	5	10.9	5	9.6
Thais kiosquiformis					5	0.5	4	0.8	5	4.8	5	2.8	5	2.7
Anadara tuberculosa					5	0.6	3	0.2	4	2.3	4	2.6	5	3.8
Anadara similis					4	0.3	3	0.2	1	0.6	4	1.4	5	2.3
Theodoxus luteofasciatus					4	4.6	3	5.3	4	6.9	2	3.4	1	0.3
Cerithidea mazatlanica					2	1.9	3	3.0	1	0.2	4	4.4	4	1.0
Mytella guyanensis					1	0.4	1	<0.1	2	0.9	2	0.3	3	0.3
Nassarius wilsoni					1	0.2			1	0.2	1	2.8	1	1.3
Protothaca asperrima					1	0.5			1	0.2	2	0.2	3	0.1
Corbula inflata					1	<0.1			2	0.2	3	0.7	4	0.9
Alpheus colombiensis					1	1.1			2	1.6	1	0.1	1	2.1
Uca saltitantia							2	1.0			2	0.5	1	0.1
Anachis nigricans							1	0.4			1	0.8	1	0.1
Anachis varia							1	<0.1			0	<0.1		
Uca brevifrons							1	0.8			0	0.1		
Uca vocator							1	1.3			0	0.2		
Gobiomorus maculatus							1	0.5						
Alpheus wickstenae									2	0.5				
Golfingia sp. 1									2	<0.1				
Nassarius luteostomum									1	0.5				
Abra palmeri									1	0.2				
Corbula obesa									1	0.2				
Corbula nasuta									1	<0.1				
Pododesmus foliatus									1	0.1				

Table 4. Continued

Zone or Subzone	Supralittoral				Upper midlittoral				Middle midlittoral				Lower midlittoral	
location	MALAGA		BUEN.		MALAGA		BUEN.		MALAGA		BUEN.		BUEN.	
species	p′	f%	p	f%	p′	f%	p′	f%	p′	f%	p′	f%	p′	f%
Axiothella sp. 1									4	2.6	3	0.8		
Loandalia sp. 1									4	0.7	1	0.1		
Protothaca grata									2	0.3	1	0.2		
Glycera abranchiata									4	8.1	5	8.6	4	11.7
Citharichthys panamensis									2	<0.1	1	0.3	5	1.1
Anadara grandis									2	<0.1	1	<0.1	3	0.2
Eleotris picta									1	<0.1	0	<0.1	1	<0.1
Citharichthys gilberti									2	<0.1	3	1.0	2	0.8
Gobionellus sagittula									1	0.2	5	6.9	5	0.7
Gobiomurus maculatus									1	0.3	2	3.1	2	2.0
Corbula ovulata											5	0.6	5	1.3
Callinectes toxotes											4	2.7	5	15.6
Batrachoides pacifici											4	2.6	5	5.1
Panopeus bermudensis											2	0.5	2	0.2
Petrolithes armatus											1	0.4	2	0.2
Palaemonetes hiltoni											1	0.3	1	2.0
Palaemon hancocki													4	1.2
Palaemon ritteri													3	0.9
number of samples	4		5		7		7		6		13		9.0	
total species richness S	10		13		34		30		45		43		33.0	
total abundance A		1911		2635		21024		30339		58874		106432		75442
abundance A′ (sta.)		2389		2635		15017		21671		49062		40935		41912
biotope diversity index H′	2.36		3.07			3.77	3.11		3.79		4.13		3.89	
equitability index J′	0.71		0.83			0.74	0.63		0.69		0.76		0.77	

Muddy flats assemblage (Table 4)

Bare soft muds of the flats in front and inside the mangroves are sometimes mixed with rocks and gravels, coarse and fine sands (whose percentages are given in Table 1), and/or enriched by wood debris and leaf accumulations. This results in local variations of the grain-size and the environmental parameters. Moisture of substrates varies depending of the tidal exposure, topography and grain-size of the sediments. Silt and clay content of these sediments is remarkably low (generally less than 20%), while either the fine sand content or the coarse sand content varies between 20 and 70%. Organic matter content is also low (generally between 2 and 4%). The resulting muddy sands or sandy muds are compacted, hard and dried at the surface, in the adlittoral and supralittoral zones of the landward areas, while they are soft (with higher water content) in lower midlittoral and upper infralittoral subzones of the outer and seaward areas. Between them, shady conditions, abundance of leefage and tidal movements keep the soil humidity high.

Upon and in these muddy substrates, epifauna and endofauna are living down to at least 30 cm deep. Excluding the insects (20 spp.), 90 species are recorded among which molluscs (gastropods: 18 spp., bivalves: 15 spp.), decapods (crabs: 15 spp., caridean shrimps: 6 spp.) and fishes (7 spp.) are the best represented.

The supralittoral zone

Muddy substrates are mainly derived from alluvial deposits. Rich in clays, they may be slightly enriched in river sands (riverine mangroves), marine sands (overwash mangroves) or gravels (fringing mangroves). Excluding the entomofauna (20 spp.), 17 species were present. This marine fauna is dominated by crabs and gastropods. Among the former, *Panopeus purpureus*, *Pachygrapsus transversus* and *Sesarma aequatoriale* (predominant) are always present. The pulmonate gastropod, *Marinula* sp. 1, is only present along the edges of the rivers where it could be abundant. Where rocks are exposed *Cerithium stercusmuscarum* has a larger distribution but with lower frequencies. Values of the diversity index are relatively high (H′ biotope=2.36 and 3.07 bit ind.$^{-1}$) in comparison to those of the canopy. Highest values are observed in stations where sediments are mixed with gravels (M8)

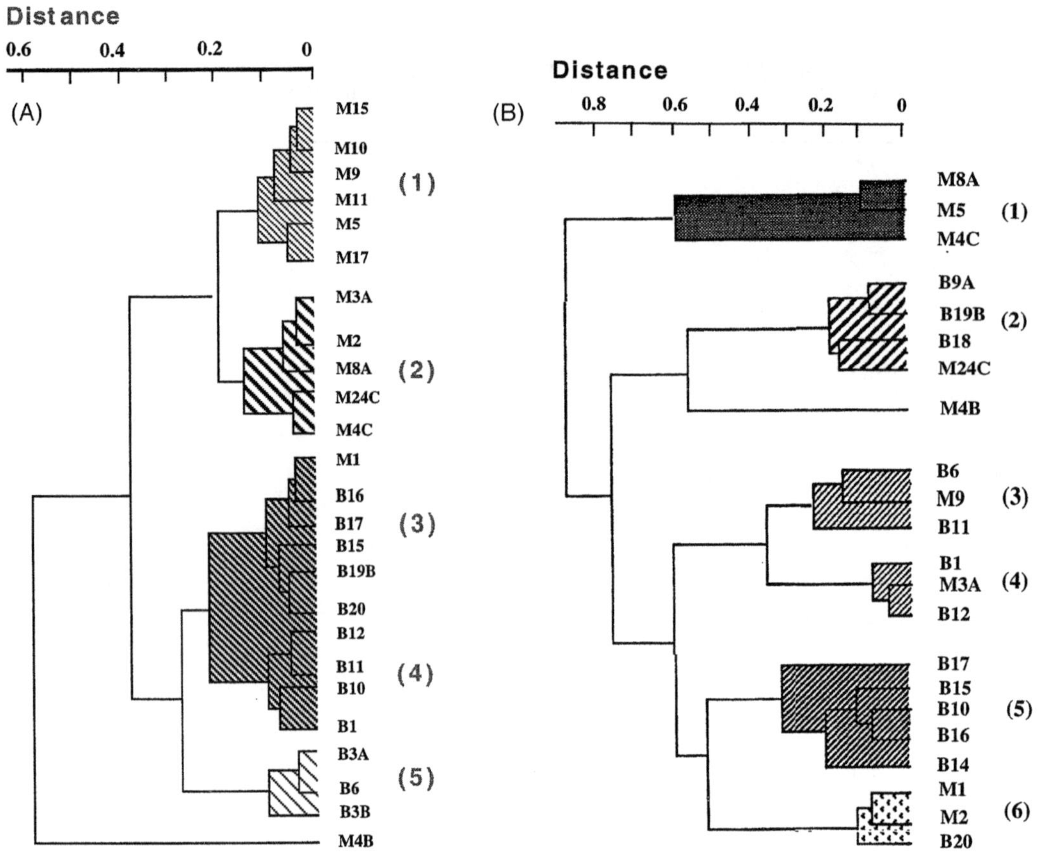

Figure 6. Clustering of samples from Buenaventura (B...) and Malaga (M...) bays according to the average distance derived from correlation index, in the supralittoral canopy (A), and in the mid-midlittoral roots (B).

or sands (B19). In the two bays, the species richness, diversity and equitability are lower landwards.

The upper midlittoral subzone

An obvious enrichment of the marine fauna is observed (46 spp.), where molluscs (17 gastropods, 9 bivalves) and decapods (13 brachyurans, 2 caridean shrimps) are the best represented. Species richness varies from 8 species (in stations of river borders) up to 25 (in the same stations that for the supralittoral zone). Commonest and dominant species are the decapods *P. purpureus* (12.4–16.2%), *Clibanarius panamensis* (14.2–19.5%), *G. pulchra* and *P. transversus*, and the gastropods *L. varia* and *Theodoxus luteofasciatus*. In the sediments, the edible *Anadara* cockles (named ‘piangua’) are common but not abundant (Betancourt & Cantera, 1976). *Talitrus* sp.1 can be very abundant under litter accumulations (up to 5000 ind m^{-2}, 12.9–35.0%) in exploited mangroves.

Species richness and diversity index are higher than in the sediments of the supralittoral zone (30–34 spp.; H′ biotope=3.8 and 3.1 bit $ind.^{-1}$). An increase of the values of both indices is observed from the entrance of the bays to their central areas (H′ sta.=1.9–3.65 bit $ind.^{-1}$ at Malaga; and 1.6–3.6 bit $ind.^{-1}$ at Buenaventura).

The middle midlittoral subzone

Aerial roots of *Rhizophora* and pneumatophora of *Avicennia* overlay most of the soil surface. In both bays, the sediments are fine muds enriched in organic matter from decaying litter. In this subzone, faunal densities increase in comparison to the upper zone and subzone, and 59 marine species (and 17 insects) were recorded, mainly represented by molluscs (gastropods: 21%; bivalves: 18%), crabs (13%), polychaetes (8%), and small fishes (8%). The most abundant species are: the hermit crab *C. panamensis* (particularly when muds are enriched in sands; 14 and 16%), the crabs

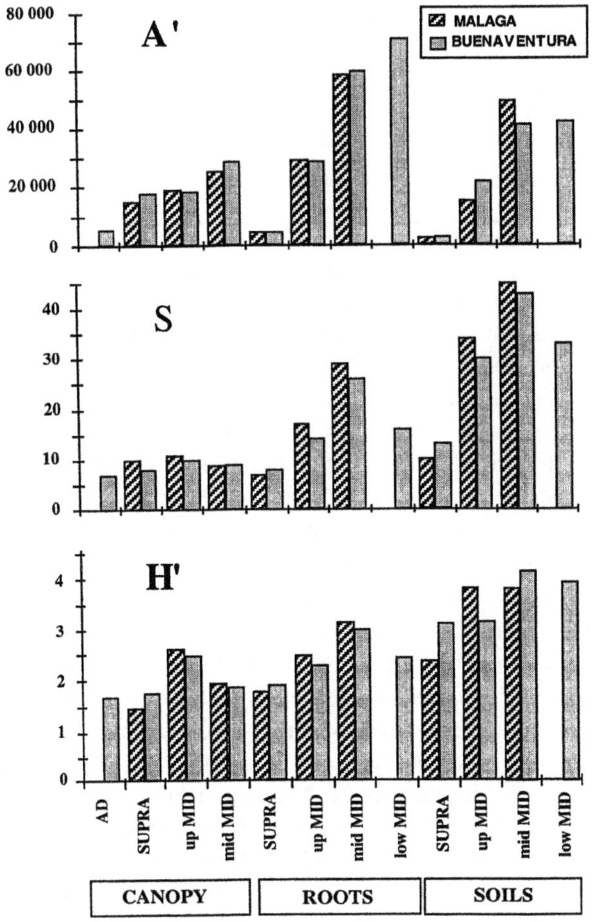

Figure 7. Variation of the relative mean abundance (A′), species richness (S) and diversity index (H′ in bit. ind.$^{-1}$) in the three biotopes of Malaga and Buenaventura bays according to zones. AD: adlittoral zone, SUPRA: surpralittoral zone, up MID: upper midlittoral zone, mid MID: middle midlittoral subzone, low MID: lower midlittoral subzone.

P. purpureus (particularly when gravels are abundant; 18.8 and 12%), *Goniopsis pulchra* (15 and 11%) and *P. transversus* (6.5 and 6%), and the polychaete *Glycera abranchiata* (8%). Where accumulations of litter occur, *Talitrus* sp. 1 is still abundant (up to 374 ind. m^{-2}), but less than in the supralittoral zone. Species richness is highly variable from one station to another (9–25 spp. at Malaga, and 9–23 spp. at Buenaventura). Mean diversity index is the highest of the midlittoral zone (H′ biotope=3.8 and 4.1 bit ind.$^{-1}$).

The lower midlittoral subzone

Along the edges of mangrove creeks, muds are more sandy and compact. Species composition and diversity are similar to those of the middle midlittoral subzone. Same species are predominant, with the addition of the swimming Portunid crab *Callinectes toxotes*. Additional species of peneid shrimps are also recorded migrating from the infralittoral zone.

Conclusion

Under the influences of strong edaphic factors, the mangal tree marine macrofauna of the Colombian Pacific coast is characterized by its low specific richness. Furthermore, it lacks some elements recorded from other mangroves, such as among the molluscs: Onchidiid pulmonate gastropods (common in South-East Asian mangroves), canopy bivalves (such as *Enigmonia rosea*) and Neritid gastropods (Cantera et al., 1983). Similarly, Macrophthalmid crabs and Periophthalmid fishes, present in eastern Atlantic (Bouchet, 1977) and Australian mangroves (Milward, 1982; Hutchings & Recher, 1982) are missing. Most species of the mangal tree Colombian fauna also live on other vertical hard substrates such as intertidal cliffs or dock pilings as recorded from other mangroves, for example by Plante (1964) and MacNae (1968).

For the faunistic analogies between the Peruvian and Colombian mangroves, see Otazu Abrill et al. (1995).

Species richness and abundance of the general macrofauna (canopy, roots and soils) appear similar in the two bays studied. But this Colombian Pacific mangrove macrofauna shows well defined assemblages related to the types of substrates (mangal substrates: canopy, trunks and roots; associated soil substrates: more or less clogged and muddy sands; litter) and to the tidal features that induce their zonation. As the specific composition of its assemblages is more dependent on the ethology of the species (tolerance to drying, feeding and reproduction...), the zonation appears more quantitative than qualitative, the same species occurring frequently in different zones or subzones.

Fine comparisons of the zonation pattern described here with that of mangroves of other regions are near impossible: there is an almost complete lack of similar comprehensive studies taking in account both the whole marine fauna (with accurate taxonomic identifications and quantitative data) and sharp zonation of its species according to tidal levels.

Canopy and branches

Among the 14 species recorded in the canopies, 11 occur in the two bays, the three other ones being uncommon *Sesarma* crabs. This assemblage shows its highest richness between the supralittoral zone and the upper midlittoral subzone. This fauna is replaced by a marine root fauna in the lowest levels of the canopy and by a terrestrial one in its highest.

So, the canopy is characterized by a poor marine fauna mixed with terrestrial species. Marine species are restricted to Grapsid crabs (*Sesarma* spp., *Goniopsis* and *Pachygrapsus*) and Littorinid gastropods, all well adapted morphologically and physiologically to hard and dried substrates (Berry, 1961; Hartnoll, 1975; Cantera et al., 1983; Reid, 1985) . Among the former that are very vagilous and vagrant, the *Sesarma* and *Goniopsis* crabs are more herbivorous (sometimes they feed on small insects in the Bromeliacean flowers) than the *Pachygrapsus* crabs that are more omnivorous (Prahl & Sanchez, 1985). In contrast, the Littorinids are herbivorous and more sedentary.

Littorina zebra and *Aratus pisoni* live in the canopy and on the roots. According to our data, the former is the dominant species in the non-perturbed mangroves, while the latter is thriving in those subjected to human impact. But *L. zebra* being consumed (under the name of 'pianqui') by inhabitants of the Pacific coast of Colombia (Ardila, 1989), its populations in the Buenaventura bay are lowered by this human impact.

Trunks and roots

Among the 35 species recorded on roots, 67% are in common to the two bays. Most of the species lacking in the Buenaventura bay are species inhabiting the gravelly muddy flats joining the mangroves (for example: the polychaetes *Neanthes galathae* and *Mystides* sp., and the gastropods *Nerita scabrocosta* and *Thais biserialis*), while those lacking in the Malaga bay live in brackish running waters. The Colombian Pacific faunistic assemblages of roots and trunks has its optimum in the middle midlittoral subzone, but these substrates can be invaded by some vagilous species of the branches (in search of more moisture) and/or by species of the soil (searching for food on the roots or among the epiphytic seaweeds).

This biotope is characterized by high density of some species. The main groups are the same as on the branches (Grapsidae and Littorinidae), associated with species that inhabit the seaweeds and the roots (sessile and boring bivalves) and by species inhabiting the nearby soft bottoms and feeding on the roots (*Talitrus* sp. 1, *Cerithidea mazatlanica*, *Anachis nigricans*, *A. varia* and *Clibanarius panamensis*). The adaptations are similar to that of the canopy species. Trophic groups include herbivores (31%), filter feeders (26%), deposit feeders (23%), and carnivores or predators (14%).

Muddy substrates with mangroves

Among the 70 species, 48% are in common to the two bays but they include all the most abundant ones. As for the roots, the species restricted to the Malaga bay are mostly associated with the gravelly mud flats, while those restricted to the Buenaventura bay are mainly associated with sandy mud flats of the river edges.

Muddy substrates with mangroves represent the upper part of the infralittoral sheltered muddy substrates due to the high water content of the sediments even when they emerge. The assemblage living under the trees is more rich in species than those of trunks and roots. It includes characteristic species of the endofauna (bivalves, shrimps, crabs and worms), colonizing sheltered shallow sediments with or without epiphytic facies such as seagrass beds or mangroves. Similar observations were made in soft bottoms of Indo-Pacific mangroves (Macnae, 1968; Day, 1974; Thomassin, 1978b). No very abundant species occur, except under litter where *Talitrus* sp. can reach densities of 5000 ind. m^{-2}, forming a temporary facies.

In conclusion, the macrofaunal assemblages associated with the Pacific Colombian mangroves formed well defined groups of species, enriched by species restricted to peculiar edaphic conditions: gravelly muddy flats and rocks at Malaga, intertidal creeks and running waters at Buenaventura. Vertical distribution of these assemblages is under the influence of the wetness and the tidal gradient. As a result, a zonation is observed without much differences between areas. This zonation appears more quantitative than qualitative and is little influenced by other factors such as physico-chemical parameters and moderate human impact.

Acknowledgements

This work was carried out under the Universidad del Valle, Cali, Colombia, and Colciencias, and analyzed in Centre d'Océanologie de Marseille (CNRS/URA no. 41), France; thanks to a grant of the European Community Commission for the senior author. The meteorological and hydrological data were provided by Himat, Colpuertos, and Ingeominas (Colombia). R. Neira contributed to the field work. Accurate identifications of the zoological material benefited of the help of the following specialists: J. Laverde-Castillo (polychaetes and sipunculids), M. Ledoyer (amphipods), C. Monniot (ascidians), C. Murillo (stomatopods), R. Neira (echinoderms), E. Pena (algae), H. Von Prahl (crustaceans and anthozoans), G. Ramos and R. Rios (carids and isopods), E. Rubio (fishes) and W. Vervoort (hydroids).

References

Alvarez-Leon, I., 1990. Mangrove ecosystems of Columbia. ITTO/ISME, PD 114: 75–111.

Ardila, C., 1989. Diagnostico de la pesca artesanal de moluscos en algunas regiones del Pacifico Colombiano. Res. Symp. int. Rec. vivos pesc., Vina del Mar, Chile: 157 pp.

Benzecri, J. P., 1973. L'analyse des données. 2. L'analyse des correspondances. Dunod, Paris: 619 p.

Berry, A. J., 1961. Some factors affecting the distribution of *Littorina saxatilis* (olivi). J. an. Ecol. 30: 27–45.

Betancourt, J. & J. Cantera, 1976. Estudio ecologico y economico de la piangua. Mem. Seminario sobre el Oceano Pacifico sudamericano (Sept 1–5 1976 Cali, Colombia). Univ. del Valle, Div. Ciencias: 475–499.

Bouchet, P., 1977. Distribution des mollusques dans les mangroves du Sénégal. Malacologia 16: 67–74.

Cantera, J. R., 1991. Etude structurale des mangroves et des peuplements macrobenthiques littoraux de deux baies du Pacifique colombien (Malaga et Buenaventura); rapport avec les conditions du milieu et les perturbations anthropiques. Thèse Doct., Univ Aix-Marseille II, Centre Océanol. Marseille: 1–429.

Cantera, J. R. & P. M. Arnaud, 1997. Structure et distribution des associations de mangrove de deux baies de la côte pacifique de Colombie. In Kjerfve, B. et al. (eds), Mangrove Ecosystem Studies in Latin America and Africa. UNESCO, Paris: 71–97.

Cantera J. R., P. M. Arnaud & B. A. Thomassin, 1983. Biogeographic and ecological remarks on Molluscan distribution in mangrove biotopes. 1. Gastropods. J. mollusc. Stud., Suppl., 12A (Proc. 2nd. Franco-British Symp. on Molluscs, Lond. Sept. 6–9, 1982): 10–26.

Chapman, V. J., 1977. Ecosystems of the world. 1. Wet coastal ecosystem. Elsevier, Amsterdam: 428 pp.

Day, J. H., 1974. The mangrove fauna of Morrumbene estuary, Mozambique. In Walsh G., S. Snedaker & H. Teas (eds), Proceedings of the International Symposium on Biology and Management of mangroves. Honolulu 2: 415–430.

Hartnoll, R. G., 1975. The Grapsidae and Ocypodidae (Decapoda: Brachyura) of Tanzania. J. Zool. 177: 305–328.

Hutchings, P. A. & H. F. Recher, 1982. The fauna of Australian mangroves. Proc. Linn. Soc. New S. Wales 106: 83–121.

MacNae, W., 1968. A general account of the fauna and flora of mangrove swamps and forest in the Indo-West Pacific region. In Yonge M. (ed.), Adv. mar. Biol. 6: 73–270.

Milward, N. E., 1982. Mangrove-dependent biota. In Clough, B. F. (ed.), Mangrove Ecosystems in Australia: Structure, Function and Management, Australian Inst. Mar.Sci., Cape Ferguson: 121–139.

Molina, A. & M. L. Mirmand, 1992. Estudio sedimentologico de la plataforma continental nordpacifica Colombiana (Bahia de Buenaventura-frontera con Panama). Bol. cient. C.I.O.H., Cartagena 10: 27–36.

Otazu Abrill, M., P. M. Arnaud & B. A. Thomassin, 1995. Les mangroves du Pérou dans le contexte côtier du Pacifique oriental: spécificités abiotiques et bionomiques. Ann. Inst. océanogr., Paris 71: 45–65.

Pérès, J. M., 1982. Zonations. In Kinne, O. (ed.), Marine Ecology, 5, Ocean management, Part 1. J. Wiley & Sons: 9–45.

Plante, R., 1964. Contribution à l'étude des peuplements de hauts niveaux sur substrats solides non récifaux dans la région de Tuléar, Madagascar. Rec.Trav. Sta. Mar. Endoume-Marseille, fasc. Hors sér., suppl. 2: 205–315.

Prahl, H. Von, J. R. Cantera & R. Contreras, 1990. Manglares y hombres del Pacifico Colombiano. Colciencias & Fondo FEN Colombia, Bogota.

Prahl, H. Von & O. F. Sanchez, 1985. Cangrejos grapsidos (Crustacea: Decapoda: Grapsidae) del Pacifico colombiano. Bol Ecotropica, Univ. Bogota 12: 31–49.

Pujos, M., R. Griboulard, C. Latouche & P. Meidinger, 1995. Analyse morpho-tectonique et caractères de la sédimentation au Quaternaire récent sur la plate-forme continentale Pacifique de la Colombie. Oceanol. Acta 18: 1–18.

Reid, D. G., 1985. Habitat and zonation patterns of Littoraria species (Gastropoda: Littorinidae) in Indo-Pacific mangrove forests. Biol. J. Linn. Soc. 26: 39–68.

Snedaker, S. C. & J. G. Snedaker (eds), 1984. The Mangrove Ecosystem: Research Methods. UNESCO, Paris, Monogr. Oceanogr. Methodology 8: 251 pp.

Thomassin, B. A., 1978a. Soft-bottom communities. In Stoddart, D. R. & R. E. Johannes (eds), Handbook of Coral Reef Research Methods, UNESCO, Paris. Monogr. Oceanogr. Method. 5: 263–298.

Thomassin, B. A., 1978b. Peuplements des sédiments coralliens de la région de Tuléar (S. W. de Madagascar) et leur insertion dans le contexte côtier Indo-Pacifique. Thèse Doct ès-Sci, Univ Aix-Marseille II, CNRS-A.O. 11581: 1-494+3 annexes & Microfiche Edition, 1983, Musée de l'Homme et Mus. nat. Hist. nat., Paris.

Vegas Vélez, M., D. Dossman & E. Rubio, 1976. Algunas observaciones oceanograficas en la Bahia de Buenaventura, Colombia. Memorias Seminario sobre el oceano Pacifico sudamericano (Sept 1–5 1976 Cali, Colombia). Univ. del Valle, Div. Ciencias: 633–661.

West, R. C., 1956. Mangrove swamps of the Pacific coast of Colombia. Ann. Assoc. Am. Geogr. 46: 98–121.

West, R. C., 1977. Tidal salt-marsh and mangal formations of Middle and South America. In Chapman, V. J. (ed.), Ecosystems of the World. 1. Wet Coastal Ecosystem. Elsevier, Amsterdam: 193–213.

Hydrobiologia **413:** 35–46, 1999.
R.S. Dodd (ed.), Diversity and Function in Mangrove Ecosystems.

Impact of ecological factors on the regeneration of mangroves in the Ciénaga Grande de Santa Marta, Colombia

Carola Elster[1,2], Laura Perdomo[3] & Marie-Luise Schnetter[1]
[1]*Botanisches Institut I, Justus-Liebig-Universität, D-35390 Giessen, Germany,*
[2]*Present address: Wiesenstr. 7, D-57290 Neunkirchen, Germany. E-mail: carolaelster@online.de*
[3]*Instituto de Investigaciones Marinas y Costeras, INVEMAR, Apartado 1016, Santa Marta, Colombia*

Key words: mangrove, regeneration, reforestation, salinity, seedlings, temperature, water supply

Abstract

In the Ciénaga Grande de Santa Marta, a brackish lagoon system on the Caribbean coast of Colombia, human-induced changes in the hydrological system caused hypersalinisation of the soils. As a result, 30 000 ha of originally 51 000 ha of the mangrove forests growing in the lagoon's flood plain died. Presently, efforts are being made to reestablish the original hydrological system. Reforestation of the bare areas is planned with the three dominating mangrove species *Avicennia germinans, Laguncularia racemosa* and *Rhizophora mangle.* The objective of the project presented here is to identify the regeneration patterns in the study area. Experiments under controlled conditions and measurements at bare and forested sites carried out from 1994 to 1996 showed that regeneration is possible. However, as the bare areas are large and presently very few or no propagules arrive naturally, regeneration will be slow. Therefore, reforestation is recommended to accelerate the process. In the forests, the seedling growth is mainly controlled by changes in salinity and the mortality by seasonal flooding, while in the bare areas, regeneration is aggravated by extremely high salinities, high soil temperatures and dessication of the top soil layers. The establishment of the small *A. germinans* and *L. racemosa* propagules is especially difficult, because they cannot establish themselves in flooded soils and in direct sunlight they dry out rapidly. The seedlings of both species are also very sensitive to changes in water level, wind, waves, temperatures exceeding 45 °C and dry soils. In *A. germinans,* phytophagous insects affect the chances for survival also. Therefore, mortality in both species is extremely high. *R. mangle* is much more resistant to most detrimental factors observed with the exception of high salinities. However, there is one favourable factor at the open sites. When sufficient water is present and salinities as well as temperatures are low enough, the high light availability increases growth rates and survival.

Introduction

Mangrove forests in arid environments tend to grow less high and exhibit a lower structural complexity compared to those in more humid areas. Descriptions of mangrove forest structures in arid zones can be found in Cintrón et al. (1978, 1980), García (1994), Serrano Días et al. (1995) and Sánchez-Páez et al. (1997). One of the main reasons for the restricted plant growth in these environments is the elevated soil salinity caused by high evaporation rates. In many cases, when flushing of the sediments is restricted, salt flats are formed. This is usually accompanied by high mortality of the trees growing nearby. Cintrón et al. (1978) discuss the possibility that mangrove areas in arid environments decrease and expand periodically in size. They attribute these variations to the influence of hurricanes and cyclic patterns in rainfall. Both, supposingly, influence the salinisation of the soils and, hence, growth and survival of mangrove stands.

The Caribbean coast of Colombia has a warm tropical climate with distinct dry and wet seasons. The Ciénaga Grande de Santa Marta (Figure 1) is located in a particularly arid zone of this coastal area. Here, evaporation is three times higher than rainfall (INDERENA, 1984) and annual precipitation rates in the north of the brackish lagoon system are only between 450 and 760 mm (Garcés & De la Zerda, 1994) increasing towards the south. Most of the scarce rain falls between August and November, the main

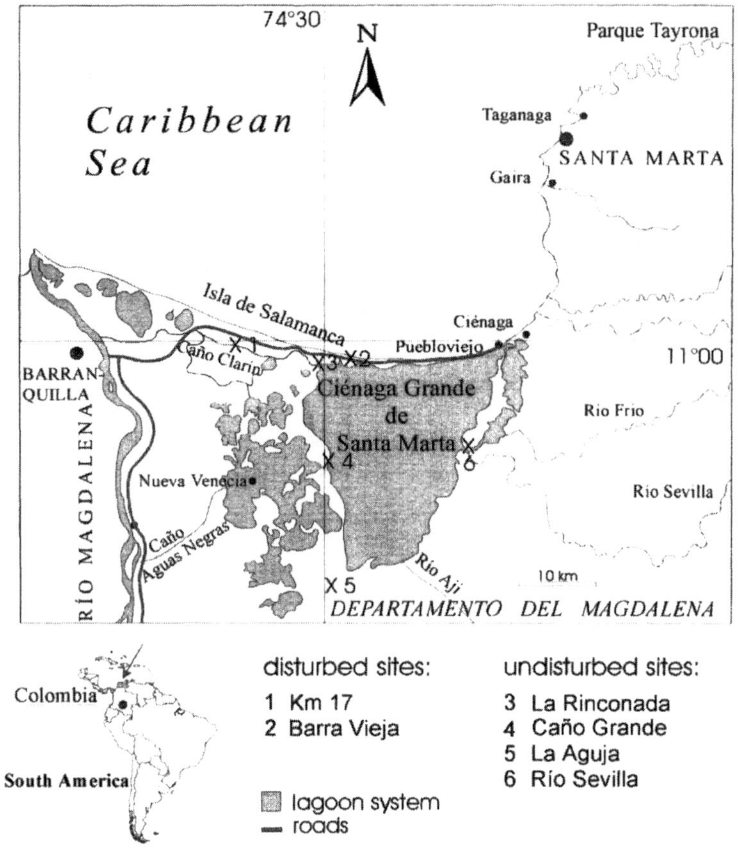

Figure 1. Location of the study sites in the Ciénaga Grande de Santa Marta in northern Colombia.

rainy season (Figure 2), when prolonged inundations occur. The hydrological system of the Ciénaga Grande is mainly influenced by the seasonal changes and by the freshwater supply from several rivers. Between the seasons, this leads to an average change of about 50 cm in the water table (Elster, 1997), while the influence of the tide is negligible. Highest tidal amplitudes are measured in the north of the lagoon system, but even here they usually do not surpass maximum values of 20 cm (Wiedemann, 1973). During the dry season (mid-December – April), drought is aggravated by strong winds, the 'brisa'. These trade winds blow predominantly from North and Northeast (Raasveldt & Tomic, 1958). The annual average temperature is about 28 °C (IGAC, 1973).

Human-induced changes in the hydrological system of the Ciénaga Grande caused a severe increase of soil and ground-water salinities. As a result, 30 000 ha of originally 51 000 ha of the mangrove forests growing in the lagoon's flood plain died (González, 1991), thereby decreasing the productivity of the ecosystem and endangering or destroying the food source for both, animals and humans in the area. The surviving mangrove stands are composed mainly of *Avicennia germinans* (L.) Stearn. In the less impacted zones, *Laguncularia racemosa* (L.) Gaertn. f. and *Rhizophora mangle* L. grow frequently also. In some open areas, the halophytes *Batis maritima* L. and *Sesuvium portulacastrum* L. thrive. Presently, efforts are being made to re-open some of the obstructed waterways to introduce more freshwater into the lagoon area. It is expected that this will lead to a re-establishment of the original hydrological system, a decrease in the salinities and a regeneration of the mangrove forests (Elster et al., 1999; Sánchez-Páez et al., 1997).

The basic objective of the project presented here was to identify regeneration patterns in the lagoonal area. We wanted to identify those ecological factors that are most influential on seedling establishment, growth and survival under the current conditions. Field work was carried out in intact forests as well as in heavily disturbed areas. The results were com-

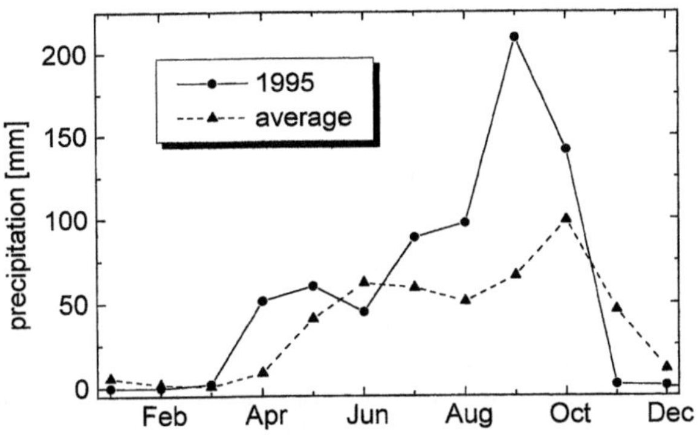

Figure 2. Precipitation rates at the airport Simon Bolivar, Santa Marta (average: 1970 – 1995; data from: IDEAM, Instituto de estudios ambientales y meteorológicos, Bogotá).

pared to identify differences between the disturbance levels.

Study sites

The present study was carried out at five intact and three highly disturbed sites well distributed in the region of the Ciénaga Grande. Study sites were chosen according to the abundance of vital seedlings and the studies were conducted near the shores of the lagoon system, where the soil was flooded during the main rainy seasons. The locations of all sites are shown in Figure 1.

Intact / forested sites

The five intact sites were: La Rinconada (10° 57' 38" N, 74° 29' 38" W), Rio Sevilla (10° 53' 11" N, 74° 19' 22" W), La Aguja (10° 40' 05" N, 74° 29' 35" W), and two sites at Caño Grande (10° 51' 33" N, 74° 28' 53" W). Here, the forests are characterized by a small belt of mainly *R. mangle* and *L. racemosa* along the shores of lagoons and channels. Further inland, the stands are dominated by *A. germinans.* Maximum tree size is about 15 – 16 m, trees are vital and produce plenty of fruit. The clayey soils are usually characterized by high contents of organic matter.

Disturbed / open sites

The three highly disturbed sites were: Kilometer 17 (Km 17) of the Caño Clarín, a channel connecting the Magdalena River with the Ciénaga Grande (10° 59' 24" N, 74° 36' 51" W), and two sites in Barra Vieja (10° 59' 20" N, 74° 26' 43" W). In earlier times, the three disturbed sites in the northwest of the lagoon system also had been covered with mangrove forests. At the time of this study, they were nearly devoid of their original vegetation.

Barra Vieja is a shallow lagoon with small islands situated on the Isla de Salamanca (Figure 1). The mangrove vegetation died about 30 yr ago, when a newly build road separated the area from the main lagoon system. In 1989, the connection between the two waterbodies was re-established and a slow natural regeneration process began. Seven years later, the young trees of *A. germinans, L. racemosa* and *R. mangle* had reached a height of only 2 m. They covered approximately one third of the sandy isles. Another third was covered by the halophytes *B. maritima* and *S. portulacastrum* and the remaining area was bare.

At Km 17, the mangrove forests had died 20 yr ago, when sediments of the Magdalena River obstructed the Caño Clarín. In 1994, when our investigations started, the efforts to re-establish the water flow had already begun. At this time, the fine muddy soils were partly flooded and dead mangrove trunks covered the zones to the left and right of the channel. Only very few stands, mostly bordering the Caño Clarín, survived. They were mainly composed of *A. germinans* and *L. racemosa* reaching a height of up to 4 m. Some *B. maritima* and *S. portulacastrum* plants grew at this site. The re-opening of the channel was finished in January 1996. In the following months, few seedlings of *A. germinans* and *L. racemosa* were able to establish themselves.

Methods

At each study site, we established a 20 m^2 plot, where the experiments were carried out monthly from June 1994 until May 1996 between 9:00 and 11:00, if not stated otherwise.

Environmental factors

Different ecological factors were measured between the investigated seedlings. If this was not possible without disturbing the young plants, we took the measurements in the seedlings' proximity.

Temperature was determined at 5 cm sediment depth, in the superficial water and in the surface water of the lagoons or channels bordering the study sites. At Km 17 and Barra Vieja, additional experiments were carried out. To identify daily changes in temperature we took measurements at 8:30, 9:30 and then every other hour until 15:30, at 10 unvegetated plots of 1 m^2 in size. In each plot, five repetitive measurements were taken at 5 cm and 0–1 cm depth as well as in the surface water (if present). The sampling at each site was carried out on three nearly cloudless days. For comparisons, only data of the same day were used to ensure similar irradiance. The plots were distinguished by their soil colour and content of sand, clay and organic matter. The measuring device was inserted directly into the soil.

The flooding height was determined by taking the inundation depth at marked reference points. In unflooded soils, the level of the water table was measured by digging a small hole into the ground and waiting for the water column to resume a constant level.

Soil moisture was determined gravimetrically by oven-drying. For each sample, wet weight was recorded followed by drying at 105° C until weight was constant (at least 24 h). Actual moisture content as a percentage of fresh soil weight was calculated. Additionally, at each site, soil was cut out without disturbing the original layering. It was put into tins with perforated bottoms and placed in a container, which was filled with water to 1 cm height (Brauner & Bukatsch, 1964). After reaching a constant weight (at least 24 h), the soil moisture was determined gravimetrically to identify the field capacity.

Salinity was examined in the laboratory with a conductivity/salinity meter (WTW, Tetra Con 91-1,5 with LF196). Ground water samples from unflooded soils were taken by digging a small hole down to the water table. A sample from the lagoons or channels bordering the study sites was taken also. When the forest floor was flooded, we determined the salinity of the surface water and took a ground water sample by inserting a plastic tube (5 cm diameter) 10 cm deep into the soil. The tube was sealed at the lower end and the lenght of the inserted section (beginning 3 cm below the soil surface) was perforated by small holes. Entry of surface water was prevented by another plastic tube fitting exactly into the first one and sealing the holes. After inserting the apparatus into the substrate we took out the inner tube allowing the ground water to leak into the outer one. To determine the salinity in the upper soil layers soil samples were taken at 2 – 10 cm depth. About 100 – 200 g of sediment were stirred into 100 ml of destilled water. The samples were left for 24 h to make sure all crystallized salt had dissolved before determining the salinity in the solution. Then, the sample was stirred again and after the sediment had settled down, the measurement was taken in the upper part of the solution, which was now clear of soil and organic matter. Later, the salinity of the original water content K_P of the soil sample was calculated by Equation (1)

$$K_P = \frac{1}{\left(1 + \frac{M^1_{H_2O}(1-K_G)}{(M^1_{H_2O}+M^2_{H_2O})K_G}\right)}, \tag{1}$$

where K_G is the measured salt concentration after dilution, $M^1{}_{H_2O}$ represents the mass of the water in the sample (difference between the weight of moist and dry soil), and $M^2{}_{H_2O}$ stands for the mass of the additionally added water. Equation (1) was derived by defining the mass of the salt in the sample, M_S, in two different ways:

$$M_S = (M^1_{H_2O} + M^2_{H_2O} + M_S)\chi K_G, \tag{2}$$

$$M_S = (M^1_{H_2O} + M_S)\chi K_P. \tag{3}$$

Then, resolving Equation (2) with respect to M_S and inserting in Equation (3) leads to Equation (1).

Establishment, growth and survival of seedlings in the field

At all intact sites, at least 10 seedlings of the three mangrove species were tagged. We only used *A. germinans* and *R. mangle* individuals that had at least one pair of leaves and *L. racemosa* plants with at least the cotyledones expanded, to make sure the period of rapid growth during the establishment and erecting phase was not included in the experiments. Total height of each seedling and number of dead plants were determined. Dead seedlings and those reaching a height of more than 25 cm were replaced by new vital seedlings. For data analysis, height differences between the months were calculated for all plants (growth rate). Propagule establishment patterns were observed also.

At the disturbed sites, similar experiments were tried three times, but only very few seedlings established themselves. Therefore, sampling could only be carried out for a limited time.

Temperature experiments under controlled conditions

Propagules for the experiments under controlled conditions were collected in La Rinconada, an intact site of the Ciénaga Grande (Figure 1). For *L. racemosa* and *R. mangle*, they were harvested directly from the trees, whereas for *A. germinans* they were collected from the ground. Propagules of all three species were set in plastic containers filled with river sand and diluted sea water. Salinity was maintained at about 32 ‰. Seedlings were left to grow until they had one or two pairs of leaves. During this time, the water level was kept at the substrate surface by watering daily with tap water. Half of the seedlings were exposed to direct sunlight and the other half was kept in the shade (10% of full irradiance).

The seedlings were dug out carefully and their roots were exposed to water of different temperatures for 30 min. Freshly collected propagules were also put into the water bath for this time span. In each group, 40 seedlings and 40 propagules were used, and the temperatures in the different treatments were kept at 30, 35, 40, 45 and 50° C. Afterwards, propagules and seedlings were set in the shade at a salinity of 32 ‰ and a waterlevel at the substrate surface. After 2 mo, the survival rate of propagules and seedlings was assessed. All propagules that had changed colour (black or brown) were considered dead. Those surviving the 2 mo had started to grow during this time. The surviving individuals in the 30° C group were used as a reference and the survival in the other groups of the same species was expressed as a percentage of the reference.

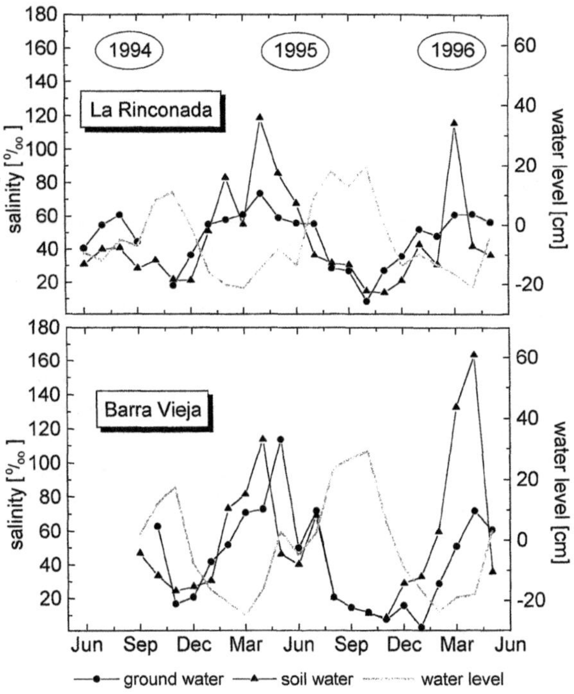

Figure 3. Salinity and water level at intact (La Rinconada) and disturbed (Barra Vieja) sites.

Results

Environmental factors

Water level

The height of the water level in the Ciénaga Grande depended on local rainfall as well as on the water input from rivers and channels. During the main rainy seasons, prolonged inundations occurred and the soil was flooded for up to three or more months. Usually, the flooding height was about 30 cm. The water level started to fall at the end of November or in the beginning of December. Lowest ground water tables were observed between January and March (Figure 3). In general, they reached about 20 cm below surface, but in extreme cases the water table was more than 50 cm below the soil surface. Intact and disturbed sites showed no differences in the fluctuation of their water level.

Soil moisture

At the intact sites with a high content of clay and organic matter, soil moisture was correlated with the height of the water level ($p < 0.0005$, Spearman Correlation, significance < 0.05). Most of the time, the actual water content was between 40% and 80%. The extreme minimum was 32% in the dry season and the maximum 90% in the rainy season. Field capacity was between 33% and 45%. The soils were at least partly, but most of the time oversaturated with water.

At the disturbed sites, soil water content was much lower. Although the very fine-grained soil at km 17 had a similar field capacity as the clayey soils in the forests (37%), the actual moisture was only between 20% in the dry season and 53% in the rainy season. Evaporation was much higher at the open sites due to the high irradiance and unflooded sediments dried up quickly. In general, the top soil layers were especially dry and at Km 17 strong trade winds eroded the very small particles. The dust settled in sheltered areas. At Barra Vieja, the soil moisture was low year round due to the low field capacity of the sandy substrates (23%). Moisture content fluctuated between 20% in the dry season and 37% in the rainy season and the fluctuations of the water level had only very limited impact on the actual water content (not significant, Spearman Correlation).

Salinity

In the forests, salinities fluctuated much less than at the disturbed sites. During the rainy season, the ground water salinity at the forested sites was about seawater concentration (36 ‰) or less. In some months, they even reached values below 10 ‰. The concentration in the ground water increased during the dry season but did not exceed 75 ‰, while the salinity of the top soil layers reached 120 ‰, in extreme cases (Figure 3). All values decreased when rainfall and flooding washed the soils.

At the disturbed sites, salinities were significantly higher ($p < 0.0005$, One way ANOVA, Kruskal Wallis, significance < 0.05). Ground water salinity ranged from 8 ‰ to 142 ‰. Even more extreme was the salinity in the top soil layers. Highest concentration observed was 221 ‰ (without figure). In some areas, salt crystallized on the soil surface, when interstitial water evaporated. The first heavy rains washed the salt out of the top soil layers into the ground water. This explains why in the ground water the salinity peak occurred a month later than in the soil (Figure 3).

Adjacent lagoons and channels usually had very low salinities. The average was between 15 ‰ and 26 ‰, while lowest values reached 0.1 ‰ and highest 37 ‰. Variation throughout the year was low.

Temperatures

At the intact sites, mean soil temperature at 5 cm depth was 29.5° C and in general, the variation was relatively small. Between 9:00 and 11:00, the lowest value observed was 25.6° C and the highest 32.7° C.

Soil temperatures were significantly higher at the open sites ($p < 0.0005$, One way ANOVA, Kruskal-Wallis) and differed much more than in the forests. Unvegetated substrates had an average temperature of 34.4° C. The lowest values were 30.2° C and the highest 40.6° C at 5 cm depth. The surface water of adjacent lagoons and channels usually had temperatures of about 32° C. In very shallow lagoons without circulation, it reached up to 40° C.

At the open sites, temperatures varied between soils of different colour and with different compounds. Dark brown clay reached the highest values. A high content of organic matter usually led to lower temperatures and white or light brown sandy soils provided the coolest substrates. In general, unflooded soils were hotter than those flooded at least some cm high (Table 1). When the water was deeper, the substrate and the surface water were heated up less. Another factor influencing these temperatures was the possibility of water exchange. When the surface water had a connection with a larger lagoon or channel, temperatures of surface water and soil were much lower. In general, highest temperatures were measured between 11:30 and 13:30, depending on wind and cloudiness.

Seedling establishment

In the forests, propagules of *A. germinans* and *L. racemosa* established mainly at the end of the rainy season, when the water table lowered. They formed thick carpets where currents washed them ashore or they got stuck in drifting plants. On the moist and shaded soils, rooting and initial growth was fast. However, the small propagules were not able to establish in inundated soils. The larger propagules of *R. mangle* did not show such distinct establishment patterns. Besides, they were able to establish in shallowly inundated soils.

At open sites, direct sun irradiation was a major problem. The small propagules of *A. germinans* and *L. racemosa* desiccated within a few days when left on the drying soils by the falling water tables. In *R. mangle,* the desiccation took approximately 2 – 3 wk. The establishment problems in flooded soils were similar to those at forested sites, but were aggravated by stronger water movements.

Seedling growth

At the forested sites, seedling growth was influenced strongly by the environmental factors. In all three species, growth was strongest at the beginning of the main rainy season, whereas it was lowest during the dry season (Figure 4). This was most obvious in *A. germinans.* The seedlings of this species grew best in September (3.2 cm/mo). During the following months, the growth rates declined steadily until the beginning of the dry season (December/January), when it reached a minimum of –1.1 cm/mo average. The negative growth rate was due mainly to herbivore attack by the caterpillars of *Junonia evarete* Cramer (Lepidoptera), which occurred only during the main rainy seasons. Drying of the tips of the seedlings was another reason for the negative growth rates. In *L. racemosa* and *R. mangle*, the tips dried out less frequently and they were rarely attacked by any herbivores. On average, *A. germinans* grew 0.6 cm/mo, *L. racemosa* 1.6 cm/mo and *R. mangle* 0.8 cm/mo.

For the growth rates of all three mangrove species, there was a significant correlation with water level (*A. germinans:* $p = 0.001$, *L. racemosa:* $p < 0.0005$, *R.*

Table 1. Temperatures in different soils depending on substrate depth and flooding (open disturbed sites; May 4,1996, 12:30; *: covered with 5 cm of surface water)

Vegetation cover	Shaded by *B. maritima*	Unvegetated			
Soil colour	Dark brown	Light brown	Dark brown	Dark brown	Dark brown
Soil compounds	High organic compound	Sand	High organic compound	Clay	High organic compound
5 cm depth	32.2° C	40.3° C	44.1° C	47.1° C	35.9° C*
1 cm depth	32.9° C	43.7° C	47.3° C	52.8° C	34.9° C*
Surface water	–	–	–	–	39.4° C*

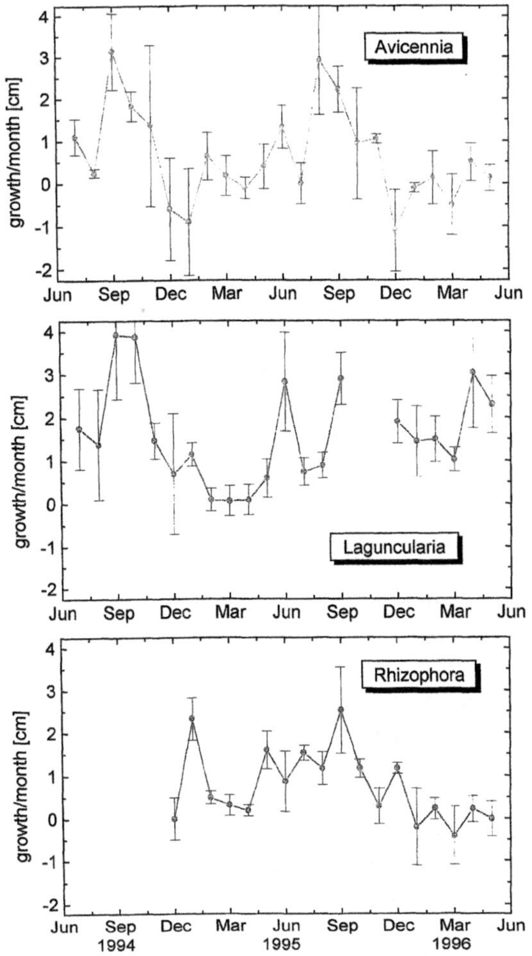

Figure 4. Growth of seedlings at intact sites (mean ± 1 SD).

mangle: $p = 0.036$, Spearman correlation) and salinity of the ground water (*A. germinans:* $p = 0.026$, *L. racemosa:* $p = 0.002$, *R. mangle:* $p = 0.027$, negative Spearman correlation). There was no correlation with the soil temperatures in the forested areas.

At the highly disturbed sites, no statistical analysis was possible due to the very limited number of surviving seedlings. It was observed though, that the growth rate was generally reduced during the dry seasons, when salinities were high. Additionally, most seedlings of *A. germinans* declined in height during the rainy season, because they were severely attacked by *J. evarete* caterpillars consuming nearly all the green plant tissue. *L. racemosa* and *R. mangle* were not seriously attacked by any herbivores and grew fast during this time.

Within 4 mo (January through May 1996), *L. racemosa* grew an average of 51.3 cm, while the tallest plant even reached 96.6 cm. In *A. germinans,* an average growth of 25.9 cm was observed and the tallest plants surpassed 50 cm. No data is available for *R. mangle,* because all tagged seedlings died during the dry season.

Seedling mortality

At the intact sites, mortality was highest in *L. racemosa* (up to 100% / mo) and lowest in *R. mangle* (maximum 36.7% / mo). Most seedlings died during the main rainy seasons (Figure 5), when water levels were highest and seedlings were submersed. In *A. germinans* and *L. racemosa,* all totally flooded seedlings died within 1 mo. In the forests, inundation was the single most important reason for mortality in these two species (*A. germinans:* $p < 0.0005$, *L. racemosa:* $p < 0.0005$, *R. mangle:* not significant, Spearman correlation). Only seedlings growing in less inundated places or taller plants were able to survive the rainy seasons.

In January 1996, another peak in mortality rates was observed in *A. germinans* and *L. racemosa,* al-

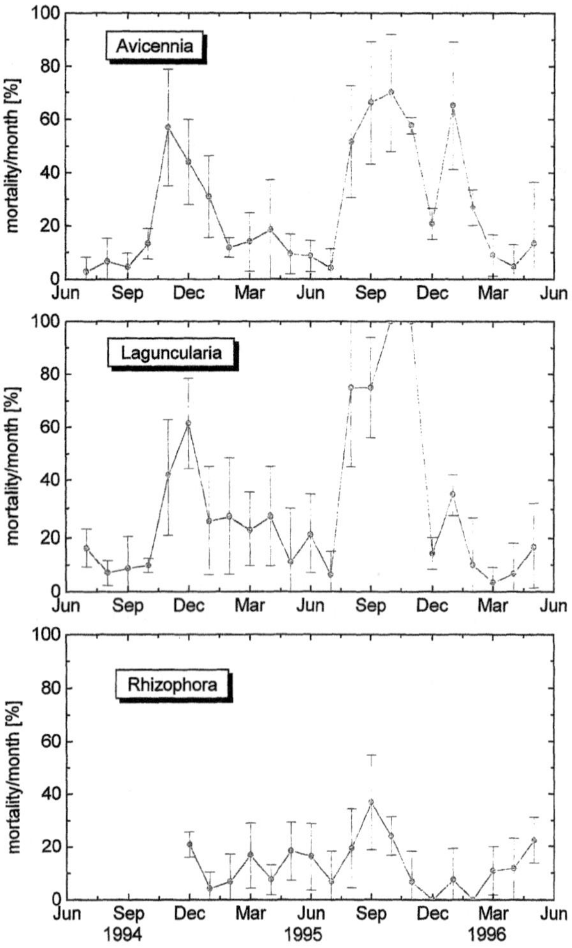

Figure 5. Mortality of seedlings at intact sites (mean ± 1 SD).

though the seedlings grew under very good conditions (low salinities, wet soil, no flooding). The high mortality was induced by the age of the plants. While older seedlings had become rare due to the high mortality during the previous months, many propagules had recently established themselves with the falling water level. Newly established seedlings, therefore, had to be included into the experiment. They had a much higher mortality rate than older plants. In *R. mangle,* no such distinct mortality and establishing phases were observed. This species was less sensitive to flooding and even totally flooded individuals were able to survive.

In the forests, low water tables (down to more than 50 cm depth) and low soil moisture during the dry seasons did not affect seedling survival in any of the three species, nor did salinity. We found a negative correlation between salinity and mortality (while salinities were low, mortality was high; *A. germinans:* $p < 0.001$, *L. racemosa:* $p < 0.001$, *R. mangle:* not significant, Spearman correlation), indicating that other factors fluctuating with the seasons, were responsible for the seedling mortality.

At the disturbed sites, mortality of all species was much higher (without figure). Especially *A. germinans* and *L. racemosa* seedlings desiccated in the dry and hot soils. Recently established seedlings died a few days after the top soil layers had dried up. *R. mangle* was the most tolerant species of drought, but most seedlings of this species died towards the end of the dry seasons due to high salinities. *L. racemosa* and, especially, *A. germinans* were more resistant to the saline substrates. Other reasons for seedling mortality at the open sites were the comparatively stronger winds and waves. The main reason for the dying of *A. germinans* was the *J. evarete* caterpillar attack. In many seedlings, nearly all the green tissue was consumed and most of the attacked plants died.

Experiments under controlled conditions

Propagules of all three mangrove species proved to be tolerant of heat. Considering *R. mangle* and *L. racemosa,* nearly 100% of the propagules survived and germinated even after being kept at a temperature of 50° C for 30 min. The least tolerant species was *A. germinans.* Only 60% of the propagules survived the 45° C and 20% the 50° C treatment (Figure 6a).

Seedlings grown in direct sun were as well relatively tolerant of high temperatures, although they were more sensitive than the propagules. *L. racemosa* and *A. germinans* survived 50° C only in very rare cases, whereas 32% of the *R. mangle* seedlings did. Most plants survived 45° C though (nearly 30% of the *L. racemosa,* 80% of the *A. germinans,* and 88% of the *R. mangle* seedlings, Figure 6b).

Seedlings grown in shade were much more sensitive to high temperatures. All of them died after the 50° C treatment and only very few seedlings of *A. germinans* and *L. racemosa* survived 45° C. *R. mangle* again proved to be the most tolerant species. 40% of the plants survived this temperature. After being kept at 40° C, most of the *A. germinans* (92%) and *L. racemosa* (60%) seedlings died, but all *R. mangle* seedlings survived this temperature treatment. *A. germinans* was even influenced by 35° C. Twenty per cent of the young plants died after the treatment (Figure 6c).

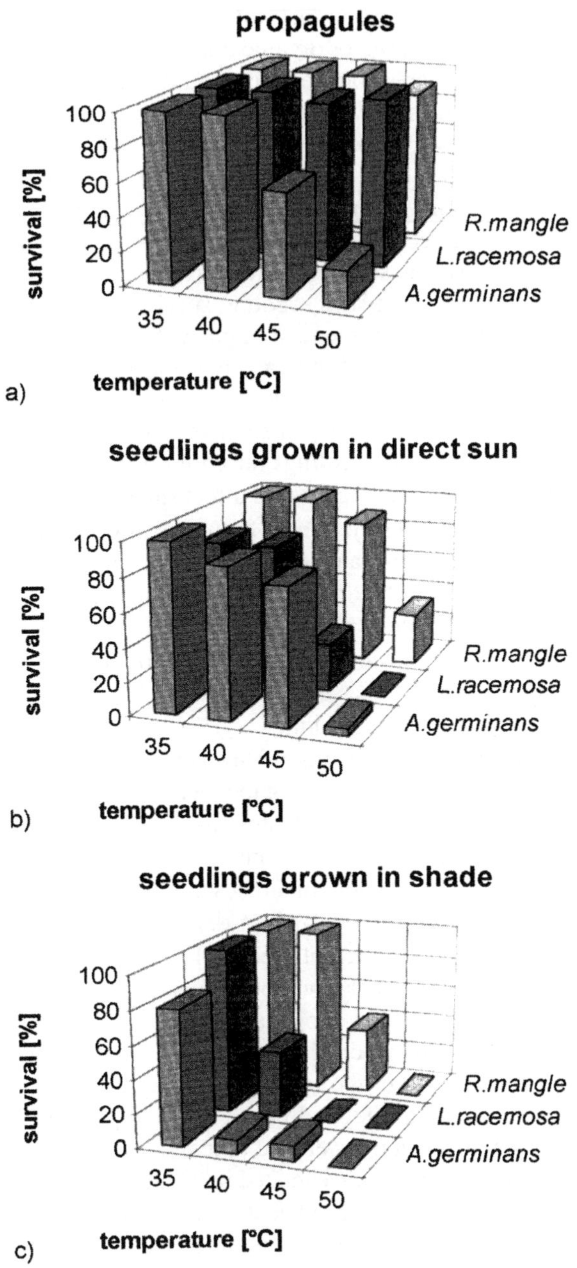

Figure 6. Impact of high temperatures on the survival of propagules and seedlings.

Discussion

Seedling survival in the intact forests of the Ciénaga Grande is influenced mainly by seasonal floodings. Neither *L. racemosa* nor *A. germinans* survive for more than 3 wk when totally submersed (Elster, 1997), therefore, many sites lose 100% of their seedlings during the rainy season due to the prolonged inundations. *R. mangle is* much more resistant to the flooding and mortality rates in general are much lower.

This indicates a major difference to the regeneration patterns in the mangrove stands in tidal areas, where prolonged inundations do not occur. But even in tidal areas, the fluctuation of the water level is considered to be one of the most important factors influencing mangrove development. Adult trees and seedlings often grow in species-specific distribution zones along the intertidal slopes (Van Speybroeck, 1992). It has been suggested that seedling development is restricted to these species-specific zones, independently of the disturbance level of the areas (Van Speybroeck, 1992). In the mangrove stands of the Ciénaga Grande, recently established *A. germinans* and *L. racemosa* seedlings experience a higher mortality than older ones, resulting in high mortality rates at the end of the rainy and the beginning of the dry season, when most propagules establish themselves. This agrees well with the mortality patterns described for other mangrove species by Tamai & Iampa (1988).

Compared to the importance of the flooding, other factors like salinity, low water tables and low soil moisture are of insignificant importance to the seedling survival in the forests. Mortality at the intact sites does not show correlations with the dissolved oxygen contents of the ground water, nor with pH, redox potentials and temperatures of the soil or ground water. Differences in light intensity between forested sites does not affect seedling development either (Elster, 1997). Herbivore attack by insects is generally low. The only impacted species is *A. germinans,* but in the forests, the attacked seedlings usually survive (Elster et al, in press). Propagule and seedling damage by crabs is an important factor influencing the regeneration in many mangrove stands (e.g. Jones, 1984; Smith, 1987a, b; Smith et al., 1989; Clarke & Myerscough, 1993; McKee, 1995). It is, however, of insignificant importance in the intact forests of the Ciénaga Grande. The only site with high abundancies of crabs is Caño Grande, but seedling establishment and survival do not differ significantly between this and other intact sites without crabs (Elster, unpublished data). Even under the conditions of the intact forest, mangrove mortality is very high and only very few seedlings survive, but due to high fruit production natural regeneration is not impeded.

In the unforested areas of the Ciénaga Grande, mangrove regeneration is aggravated by several other factors: even sites influenced by fresh water after the re-opening of obstructed channels exhibit high salinities during the dry seasons. During this time, high evaporation rates foster the rising of very saline ground water and the formation of salt crusts. The direct solar radiation additionally leads to very high temperatures and extreme drought in the top soil layers. Thus, in wide parts of the disturbed areas natural regeneration does not occur.

The heat itself does not seem to make propagule and seedling establishment and survival impossible, as long as sufficient water is present. Our temperature experiments show that all studied mangrove species, but especially *R. mangle,* belong to the more temperature-resistant plants. Larcher (1984) defines those species as heat sensitive that in a heat-adapted state tolerate only 30 min in 30–40° C. Forty-five °C is the extreme limit for their survival. More robust species resist 30 min in 50–60° C. The limit for highly differentiated plant cells seems to be at 60–70° C. The 30-min exposure time used normally does not show the real temperature limits for a given species, though. Extending the treatment time to an hour lowers the resistance level by 1–2° C (Larcher, 1984). The exact heat tolerance of mangroves being kept at high temperatures for several hours as they occur in the Ciénaga Grande is not known yet. McMillan (1971) showed that rooted but stemless *A. germinans* seedlings survive 10-min exposures to 43° C, but 48-h exposures cause death and decay. In his experiment, rooted seedlings with stems and leaves were more heat resistant than stemless individuals. It is not clear though whether his results are applicable to other *A. germinans* populations since he used propagules from temperate plants that seldom exceeded 1.5 m height. Populations growing in the tropics may be much more heat resistant.

In our experiments, propagules of *R. mangle* and *L. racemosa* proved to be very tolerant of heat. Since they are not affected at all by a 50° C treatment, most of them will survive the extreme temperatures in the open areas. Even on sunny days, temperatures in most wet soils do not exceed 50 – 53° C at less than 1 cm depth. However, higher temperatures are recorded for the unvegetated areas of the Ciénaga Grande. The reported maximum is 56.9° C (Ensminger, 1996). Such extreme values are only reached in unflooded and relatively dry soils. At overheated sites, propagules of *A. germinans,* will have a much lower survival rate than the other two species. *A. germinans* propagules will probably not survive the extreme heat in the top soil layers of dark substrates, but it can colonize more favourable places like moist soils of a light colour with high content of sand or organic matter.

Seedlings are more sensitive to high temperatures than propagules. However, most sun-adapted individuals survive 45° C in a water bath. In the Ciénaga Grande, soil temperatures of 47° C are reached at a depth of 5 cm, but normally values stay much lower. Most of the seedlings roots grow at this depth. They are the most heat sensitive plant parts (Larcher, 1984). Therefore, at most sites seedling survival is possible. There will be a problem, though, with reforestation measures, when seedlings are not adapted to direct sun and the co-occuring high soil temperatures. Most seedlings will die during their first sunny day in the new environment. Seedlings raised for transplantation should, therefore, be grown in direct sun or at least be acclimatized to it some weeks before planting. If this is not possible, transplants should be shaded in the field. The shade should then be reduced gradually. There is a chance that older plants (1 yr or more), which are generally less sensitive, might be more tolerant of high temperatures and that they might survive to a higher percentage than the seedlings.

The drought of the top soil layers is one of the most important factors influencing the regeneration of mangroves in the unvegetated sites. Especially the small propagules of *A. germinans* and *L. racemosa* are very sensitive to desiccation. In direct sunlight, they usually dry out and die in 2 – 4 d on dry sand. *R. mangle,* again, is the most tolerant species. The propagules can survive several weeks without water (Elster, 1997). However, in no case propagules will establish themselves when lying on dry soil. Some of them start growing roots, but the roots dry out rapidly and die. Recently established seedlings, too, die rapidly. Only plants with a better developed root system can survive the drought. Most unvegetated soils are dried out as well as overheated. Both factors feed back on each other and in many cases their impact on propagules and seedlings cannot be distinguished.

High salinities are well known to be one of the major factors influencing mangrove development (e.g. Cintrón et al., 1978, 1980; Clough, 1984; Tomlinson, 1986; Ball, 1988). It is the single most important factor for the mangrove mass mortality of the past years in the Ciénaga Grande (Elster et al., 1999, González, 1991). In the intact forests of this area, salinity only influences mangrove growth rates, but it is a major factor influencing seedling survival and growth at the disturbed sites. Especially the high salinities occurring at the end of the dry season are detrimental to mangrove seedlings and nearly all of them die.

Although some of the factors discussed above may not exceed values detrimental for the individual seedlings, they at least weaken the young plants. Due to mutual reinforcement, the seedlings die eventually even in those areas, where each single factor may be below the survival limit for mangroves.

Seedling survival and growth at open sites is under otherwise favourable conditions (wet soils, low salinities, moderate temperatures) much higher than in forests, which is due to higher light intensities. In the forests of the Ciénaga Grande, mean values of relative light intensity reach about 22% (Elster, 1997), which is much higher than in mangrove forest of southern Thailand, where light intensities of 6.1% and 10.9% are observed (Tamai & Iampa, 1988). At the open sites of the Ciénaga Grande, recently established *A. germinans* seedlings grow about 6.5 cm/mo (January 1996 through May 1996), which is more than nine times as much as found under intact canopies (0.7 cm/mo) during the same time. In *L. racemosa,* the growth increases for more than 6.7 fould (1.9 cm/mo in the forests, 12.8 cm/mo at the open sites). Similarly high growth differences are reported for seedlings of *Rhizophora apiculata* Blume, *Bruguiera parviflora* Wright & Arn. and *B. cylindrica* Blume. Elongation rates in these species are up to six times as high under full light intensity as under intact canopies. In young trees, individuals of these species grow more than 10 times as fast at open sites (Tamai & Iampa, 1988). Open areas with favourable soil conditions, therefore, can experience a rapid regeneration.

Disturbed sites with a recently reestablished hydrological system, low salinities and sufficient water are such favourable areas. Rapid colonization and seedling growth of all three mangrove species can be observed where their propagules are present. In large parts of the open areas, however, no or only very few propagules arrive naturally (Elster, pers. obs.). This is also described by Ensminger (1996), who found that the disturbed sites close to the Caño Clarín are only reached by propagules of *A. germinans* and only very few or no propagules of *L. racemosa* and *R. mangle* arrive. This is mainly due to the survival of *A. germinans* stands close to the channel.

The various species of the genus *Avicennia* are generally considered to be the ones most tolerant of high salinities (Clough, 1984; Ball, 1988; Elster, 1997). This is the main reason why *A. germinans* had the highest survival rate of all mangrove species in the Ciénaga Grande during the past years. Some *A. germinans* stands even survived at the shores of the otherwise totally degraded northern part of the lagoon system (Elster, pers. obs.). After reopening the Caño Clarín, these stands started to flower and set fruit (Ensminger, 1996) and short distance dispersal led to an increase of *A. germinans* propagules at the highly disturbed sites next to the channel. Obviously, long distance dispersal is limited in the Ciénaga Grande, since the number of propagules decreases rapidly with increasing distance to the fruiting trees. These observations agree well with the results of Tamai & Iampa (1988), who report a similar dispersal pattern for mangroves in southern Thailand. Many disturbed sites of the Ciénaga Grande are not reached by propagules naturally and, therefore, regeneration will be very slow in these completely devastated parts of the lagoon system, even if salinities will decrease.

Reforestation can accellerate the regeneration process. *A. germinans* would be the appropriate species to colonize disturbed areas due to its tolerance of high salinities, whereas in many sites *L. racemosa* and *R. mangle* will experience severe problems with the increasing salinities during the dry seasons. On the other hand, *A. germinans* seedlings will have great difficulties to establish themselves due to the strong herbivore attack by *J. evarete* caterpillars. As the abundance of caterpillars is very low during the dry season, this species should be planted at the beginning of January to reduce chances for caterpillar attack (Elster et al., in press).

At sites with low soil salinities, the establishment of *R. mangle* would be very successful. This species is more resistant to most detrimental factors observed in the Ciénaga Grande than *A. germinans* and *L. racemosa.* The size of the *R. mangle* propagules provides many advantages that can be used for reforestation. The propagules can be set with their lower end 3 – 4 cm deep into the ground and will neither be harmed by inundation or currents and waves of the surface water, if present, nor by insects, heat or short drought phases. The small propagules of *A. germinans* and *L. racemosa,* on the other hand, cannot establish themselves in inundated soils, instead, they will be carried away by currents. On dry soils, they will desiccate rapidly. Also their seedlings are much more delicate. They are easily damaged by wind, waves and drought.

Acknowledgements

We thank INVEMAR and CORPAMAG/PRO-CIÉNAGA (both Santa Marta, Colombia) for the collaboration and the access to laboratory as well as boat facilities. Special thanks are due to A. Peter for constructive criticism on the manuscript, as well as to C. Carbono and E. Cabas for their assistance with the field work. O. David (OCENSA) provided the precipitation data, and L. M. Constantino (Fundación Herencia Verde) helped with insect identification. The research was supported by the GTZ (TÖB, "Flanking Program of Tropical Ecology", PN: 90.2136.1), Eschborn, Germany.

References

Ball, M. C., 1988. Salinity tolerance in the mangroves *Aegiceras corniculatum* and *Avicennia marina* I. Water use in relation to growth, carbon partitioning and salt balance. Aust. J. Plant Physiol. 15: 447–464.

Brauner, L. & F. Bukatsch, 1964. Das kleine pflanzenphysiologische Praktikum. 7. Auflage, Gustav Fischer Verlag, Stuttgart, Germany: 288 pp.

Cintrón, G., C. Goenaga & A. E. Lugo, 1980. Observaciones sobre el desarrollo del manglar en costas áridas. Memorias del Seminario sobre el estudio científico e impacto humano en el ecosistema de manglar, Cali, 1978. UNESCO: 18–32.

Cintrón, G., A. E. Lugo, D. J. Pool & G. Morris, 1978. Mangroves of arid environments in Puerto Rico and adjacent islands. Biotropica 10: 110–121.

Clarke, P. J. & P. J. Myerscough, 1993. The intertidal distribution of the grey mangrove *(Avicennia marina)* in southern Australia: the effects of physical conditions, interspecific competition and predation on propagule establishment and survival. Aust. J. Ecol. 18: 307–315.

Clough, B. F., 1984. Growth and salt balance of the mangroves *Avicennia marina* (Forsk.) Vierh. and *Rhizophora stylosa* Griff. in relation to salinity. Aust. J. Plant Physiol. 11: 419–430.

Elster, C., 1997. Beziehung zwischen ökologischen Faktoren und der Regeneration dreier Mangrovenarten im Gebiet der Ciénaga Grande de Santa Marta, Kolumbien. Dissertation, Justus-Liebig-Universität, Giessen: 220 pp.

Elster, C., J. Polanía & O. Casas-Monroy, 1999. Restoration of the Magdalena River Delta, Colombia. Vida Silvestre Neotropical 7: 23–29.

Elster, C., L. Perdomo, J. Polanía & M.-L Schnetter, 1999. Control of *Avicennia germinans* recruitment and survival by *Junonia evarete* larvae in a disturbed mangrove forest in Colombia. J. Trop. Ecol. 15 (in press).

Ensminger, I., 1996. Hydrologische Veränderungen am Canal Clarín und ihre Bedeutung für die Regeneration salzgeschädigter Mangrove. Diploma thesis, Justus-Liebig-Universität, Giessen: 137 pp.

Garcés, G., D. M. & S. De la Zerda L., 1994. Gran Libro de los Parques Nacionales de Colombia. Circulo de lectores S.A. Bogotá: 230 pp.

García, C. B., 1994. Quantitative data on the forest structure of a small mangrove formation in the bay of Nenguange, Colombian Caribbean. An. Inst. Invest. Mar. Punta Betín 23: 183–187.

González, E., 1991. El manglar de la Ciénaga Grande de Santa Marta: Ecosistema en peligro de extinción. Colombia, sus gentes y regiones. Instituto Geográfico 'Agustín Codazzi', GAC 21: 2–21.

IGAC, 1973. Monografía del departamento del Magdalena. Instituto Geográfico 'Agustín Codazzi', Bogotá: 162 pp.

INDERENA – Divisiones fauna terrestres y parques nacionales, 1984. Colombia Parques Nacionales. INDERENA, Bogotá: 264 pp.

Jones, D. A., 1984. Crabs of the mangal ecosystem. In Por, F. D. & I. Dor (eds), Hydrobiology of the Mangal. Dr W. Junk Publishers, The Hague: 89–109.

Larcher, W., 1984. Ökologie der Pflanzen auf physiologischer Grundlage. 4. überarbeitete Auflage, Verlag Eugen Ulmer, Stuttgart: 403 pp.

McKee, K. L., 1995. Seedling recruitment patterns in a Belizean mangrove forest: effects of establishment ability and physico-chemical factors. Oecologia 101: 448–460.

McMillan, C., 1971. Environmental factors affecting seedling establishment of the black mangrove on the central Texas coast. Ecology 52: 927–930.

Raasveldt, H. C. & A. Tomic, 1958. Lagunas Colombianas. Revista de la Academia Colombiana de Ciencias Exactas, Físicas y Naturales, Bogotá 10(40): 175–198.

Sánchez-Páez, H., R. Alvarez-León, F. Pinto-Nolla, A. S. Sánchez-Alférez, J. C. Pino-Renjifo, I. García-Hansen & M. T. Acosta-Peñaloza, 1997. Departamento del Magdalena. In Sánchez-Páez, H. & R. Alvarez-León (eds), Diagnostico y Zonificación Preliminar de los Manglares del Caribe de Colombia. Ministerio del Medio Ambiente, Bogotá: 119–191.

Serrano Días, L. A., L. Botero, P. Cardona & J. E. Mancera-Pineda, 1995. Estructura del manglar en el delta exterior del Río Magdalena-Ciénaga Grande de Santa Marta, una zona tensionada por alteraciones del equilibrio hídrico. An. Inst. Invest. Mar. Punta Betín 24: 135–164.

Smith, T. J. III, 1987a. Seed predation in relation to the dominance and distribution in mangrove forests. Ecology 68: 266–273.

Smith, T. J. III, 1987b. Effects of seed predators and light level on the distribution of *Avicennia marina* (Forsk.) Vierh. in tropical, tidal forests. Estuar. coast shelf Sci. 25: 43–51.

Smith, T. J. III, H. T. Chan, C. C. McIvor & M. B. Robblee, 1989. Comparison of seed predation in tropical tidal forests from three continents. Ecology 70: 146–151.

Tamai, S. & P. Iampa, 1988. Establishment and growth of mangrove seedlings in mangrove forests of southern Thailand. Ecol. Res. 3: 227–238.

Tomlinson, P. B., 1986. The botany of mangroves. Cambridge University Press, Cambridge: 419 pp.

Van Speybroeck, D., 1992. Regeneration strategy of mangroves along the Kenya coast: a first approach. Hydrobiologia 247: 243–251.

Wiedemann, H. U., 1973. Reconnaissance of the Ciénaga Grande de Santa Marta, Colombia: physical parameters and geological history. Mitt. Inst. Colombo-Alemán Invest. Cient. 7: 85–119.

Hydrobiologia **413:** 47–52, 1999.
R.S. Dodd (ed.), Diversity and Function in Mangrove Ecosystems.

Mangrove rehabilitation: choice and necessity

C. D. Field
Faculty of Science (Gore Hill), University of Technology, Sydney. P.O. Box 123, Broadway NSW 2007, Australia

Key words: mangrove, rehabilitation, conservation, sustainable production, coastal protection

Abstract

The concept of mangrove ecosystem rehabilitation is considered. Four main reasons for rehabilitating mangroves are identified: conservation, landscaping, sustainable production and coastal protection. Practical aspects of mangrove rehabilitation, such as the causes of site degradation, site selection, source of seedlings and planting, monitoring and maintenance are then briefly mentioned. Future developments that may impact on mangrove ecosystem restoration such as the importance of biodiversity, biotechnology, ecological modelling, mapping, human ecology and data bases, are then briefly reviewed. Finally, the matter of choice and necessity is addressed.

Introduction

In recent years, the pressures of increasing population, food production, industrial and urban development and wood chipping have caused a significant proportion of the world's mangrove resource to be destroyed. There is a necessity to rehabilitate mangrove lands to support human communities. This may involve a landscape mosaic composed of mature forests, logged forests, aquaculture and agriculture. The use of such a mosaic can be sustainable.

Another impetus behind the rehabilitation of mangrove ecosystems is the spectacular rise of environmental consciousness over the past 30 years. The rehabilitation of mangrove lands for conservation remains a matter of choice. Such a choice should be determined by the socio-economic priorities of the communities involved but it is often influenced by pressure from conservation organisations advocating preservation of nature.

The concept of rehabilitation of a mangrove ecosystem

The continuum in the changes that mangrove lands can undergo is well illustrated (Figure 1) by a modified diagram from Brown & Lugo (1994). *Rehabilitation* is the return of degraded mangrove land to a fully functional mangrove ecosystem regardless of the original state of the degraded land. *Restoration* is the return of degraded mangrove land to something like its presumed original state.

Objectives of rehabilitation

Rehabilitation is often required because of intensive use of the sites, excessive extraction of materials, chemical contamination and changes to the hydrology of the site; though at times it can arise because of climatic impacts that have destroyed the natural vegetation. It is essential to define objectives as a first step in the rehabilitation process. The question that must be addressed is what is required from the final outcome of the rehabilitation process. Definition of the objectives will identify a clear framework for implementation, operation and appraisal of success.

Four main criteria measure the success of a mangrove rehabilitation programme. These are the effectiveness of the planting, which can be considered as the closeness to which the new mangrove ecosystem meets the original objectives of the rehabilitation programme; the rate of recruitment of flora and fauna, which can be considered a measure of how quickly the rehabilitated site recovers its integrity; the efficiency of rehabilitation, which can be measured in terms of the amount of labour, resources and material used and the long term sustainability of the new mangrove ecosystem. The effectiveness and efficiency are only sometimes quantified, the recruitment of flora and fauna rarely quantified and the long term sustainability awaits determination.

Approaches to mangrove rehabilitation

There are four main reasons for mangrove ecosystem rehabilitation and the degree to which the original ecosystem is rehabilitated may vary in each case.

Conservation

If a degraded mangrove ecosystem is being rehabilitated for conservation purposes, all ecological processes and genetic diversity must be reinstated. This form of *rehabilitation* equates to *restoration*. A number of factors may influence the similarity of the restored mangrove ecosystem with any mangrove ecosystem that may have previously occupied the site such as irreversible changes to the site, natural variability of the mangrove ecosystem and lack of bench mark data for the original mangrove ecosystem. In mangrove rehabilitation of this type, the subsequent level of practical management is often very low and quantification of the success of the rehabilitation rarely goes much beyond assessment of the growth of the trees. There are few examples of this type of mangrove rehabilitation (Field, 1996)

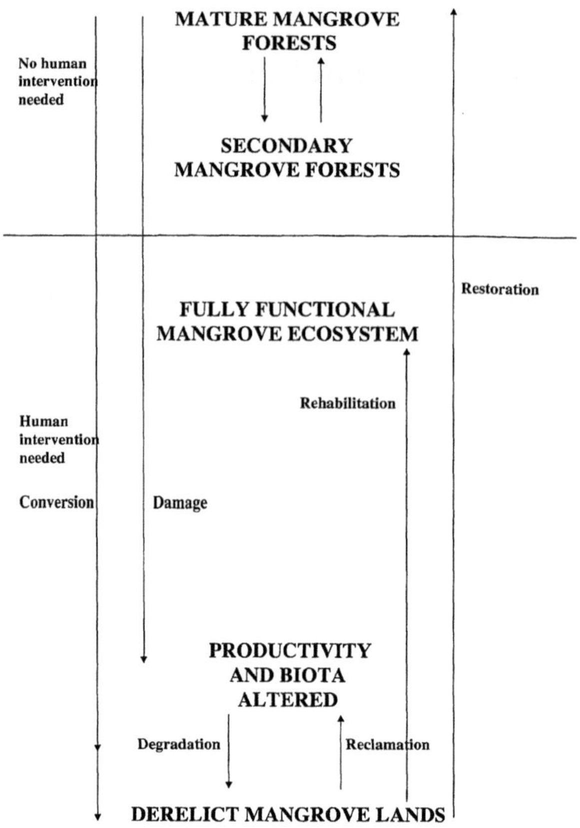

Figure 1. Common terms used to discuss changes that mangrove lands can undergo (from Brown and Lugo 1994).

Landscaping

The use of mangroves for the purpose of enhancing a landscape is a special example of rehabilitation for conservation. However, in this case, there is usually no attempt made to recreate a previously existing mangrove ecosystem; rather the goal is to produce a sustainable mangrove ecosystem that will enhance a coastline or estuary. Landscaping of arid coastal regions using *Avicennia marina* has been attempted around the Arabian Peninsula. In Australia, there are examples of mangroves being planted as part of a resort complex.

Multiple use systems for high and sustainable yield

Mangrove ecosystems can be managed as multiple use systems for the high and sustainable yield of natural products. This implies careful management and perturbation of the ecosystem without loss of productivity. Examples of the use of mangrove ecosystems for sustainable yield of natural products are timber and charcoal production (Chan, 1996) and shrimp production (Robertson & Phillips, 1995). Unfortunately, many of the attempts to utilise mangrove ecosystems in this way have ended in disaster as a result of poor, short-term and greedy management practices. However, this should not lead to the conclusion that mangrove ecosystems cannot be managed to deliver high yields of natural products on a sustainable basis. Sustainable production of natural products from mangrove ecosystems must seek to avoid environmental disasters in the short and long term and to encourage preservation of the natural system as much as possible. In such endeavours, there are conflicting goals to be considered. These are the preservation of environmental integrity, economic efficiency and equity for the local community.

This type of rehabilitation seeks to regain the fertility of the land without undue regard to how the rehabilitated mangrove ecosystem compares with any previous one. The main objective is to increase productivity.

Protection of coastal areas

The planting of mangroves along coastlines damaged by cyclones and tidal bores occurs in countries such as Vietnam, China and Bangladesh. In Bangladesh (Saenger & Siddiqi, 1993) 120 000 ha of mangroves have been planted since 1966. Nowhere else have mangroves been planted on such a large scale. In this case, the mangroves were planted on newly accreted land. Two species of mangrove, *Sonneratia apetala* and *Avicennia officinalis*, dominate the mangrove plantations, usually as monospecific stands. The planting of mangroves has been highly successful in protecting and stabilising coastal areas and in providing substantial timber production. Such artificially constructed mangrove forests seem highly beneficial but little attempt has been made to study their ecology.

Specific considerations when rehabilitating mangroves

There are five main practical considerations when contemplating the rehabilitation of a mangrove ecosystem. They are: identifying the causes of site degradation, assessing site selection criteria, sourcing of seedlings and planting (natural, field transplant or nursery), monitoring the outcomes and maintaining the resulting mangrove ecosystem .

There are well established general criteria in the literature for these activities (Lewis, 1994; Field, 1996; JAM, 1997) however local environmental conditions must dominate all other considerations.

The future of mangrove rehabilitation

If one believes the fashionable trends in science then one can expect a rapidly rising world population, an acute shortage of land for crops, a shortage of potable water, dwindling fish stocks and sea-level rise, due to climate change. All these outcomes may affect the future of mangrove ecosystems and will probably accelerate the present decline in their global areal coverage.

Mangrove research has concentrated on understanding the structure and function of pristine mangrove ecosystems though there is an emerging literature on more applied problems. There remains a good case for the study of pristine mangrove ecosystems but more scientific effort needs to be directed to the optimal use of the mangrove ecosystem.

Six activities underpinning the future direction of rehabilitation will be examined briefly.

Biodiversity

As a means of maintaining global biodiversity increasing attention is being given to restoring and creating various coastal habitats. This activity is expensive and populist and probably involves more mangrove research dollars then any other research activity.

The conservation of biological diversity is central to dogma of the international conservation community. It is perceived as pivotal to nature conservation because species extinction threatens not only the idealised Western view of nature but depletes the genetic resources that are essential for continued human prosperity (Davie & Hynes, 1997). However, it is interesting to note that the relationship between changes in biodiversity and ecosystem function is not easily quantified in mangrove ecosystems despite the extensive pool of information (Twilley et al., 1996). In addition, the value of focusing the purpose of nature conservation on biodiversity can be queried. Davie & Hynes (1997) question why biodiversity should be paramount in the thinking of conservationists and argue that conservation should be more inclusive of community participation. They believe that nature conservation practice should stand within a context of multiple land use. They argue that the conservation of ecological processes to maintain habitat, water quality and soil fertility may better integrate nature conservation into other land uses.

Perhaps of even more importance is the relationship between biodiversity and productivity. There is some evidence that productivity and biodiversity have a reciprocal relationship (Johnson et al., 1996). However, biodiversity and productivity both contribute to system integrity, which is pivotal to management. The question that needs to be asked with respect to mangrove ecosystems is how much loss of biodiversity can be tolerated in order to maximize productivity and ensure system integrity. There are plenty of highly managed, relatively stable, highly simplified, 'natural' environments that support human populations.

Biotechnology

Biotechnology is having an impact in the plant sciences and it is now possible to design plants with resistance to insects, heavy metal toxicity, herbicides and environmental stress. Genes modifying wood properties and other plant traits can be engineered into trees. The production of genetically improved trees for forestry plantation will improve yields and growth rates. The technology for tissue culture has now been refined so that mass production is cheap and reliable.

Mangroves are beginning to be investigated using the techniques of molecular biology and tissue culture. The genetic characterisation of a species and the degree of polymorphism is basic to any understanding of variation within mangrove ecosystems. It seems that it is difficult to partition phenotypic variability in mangrove species into environmental and heritable components (Lakshmi et al., 1997). The molecular basis for salt tolerance in mangroves is another area under investigation.

Forest tree breeding involves the selection of superior germplasm and the large scale reproduction of elite trees. In the case of mangroves, morphological and physiological markers can be used to identify superior trees. Elite trees can then be produced using micropropagation, vegetative propagation or seed propagation.

The ability to select mangroves for particular sites and to engineer the quality and mass of the wood produced is now in the realms of possibility. However, at this time no transgenic mangrove has been regenerated.

There is no doubt that the technology exists to shape the future of mangrove rehabilitation. The dilemma is that given the relatively small world-wide area of mangroves and their lack of economic significance globally, it is unlikely that sufficient resources will be put into mangrove biotechnology to ensure that the procedures will become wide-spread.

Modelling of ecological processes in mangroves

There are many mangrove rehabilitation projects with various aims that have been undertaken in the last few years or that are currently underway (Field, 1996; JAM, 1997). In the same period, there has been an explosion of scientific papers on mangrove biology and ecology. Mangrove ecologists tend to be primarily concerned with the intrinsic nature of their research rather than in initiating the use of their findings in the management of mangrove rehabilitation projects. There is a paucity of ecological studies on heavily degraded mangrove ecosystems and little attempt to extrapolate ecological findings from normally functioning mangrove ecosystems to those existing under stressed conditions.

However, recently there has been an attempt to model the dynamics of mangrove forest development (Chen & Twilley, 1998). The model simulates mangrove forest development by integrating demographic processes with soil nutrient availability. The results from the model against measured data provide strong support for the dynamics of succession predicted by the model. In a subsequent paper (Twilley et al., 1998) the model has been extended to establish trajectories of mangrove response in rehabilitation projects. This should assist in the design and implementation of such projects.

If the model can be extended to simulate mangrove rehabilitation projects using existing data sets and so provide useful information for land use managers, then ecology and management will begin to work together. Underwood (1995) points out that ecologists should not be too eager to confine their efforts solely to the provision of sound ecological advice but should be prepared to have more say in the way the advice and data are used.

There is a need for applied ecological research aimed at testing the decisions made when rehabilitating mangroves. This seldom happens. Indeed, much mangrove research is done in isolation from the needs of the managers of the rehabilitation projects. There is an urgent need to study the failures of mangrove ecosystem management. Likewise, there is a need to do research that will enhance mangrove ecosystem rehabilitation.

The modelling of mangrove ecosystems has the merit of trying to utilize the vast amount of information on mangroves and to provide guidance to land use managers. It is not a high cost exercize but one that has the potential to provide useful results.

Geographical Information Systems (GIS) and mapping

If the global extent of mangroves is to be monitored, then a base-line has to be established against which future trends can be measured. A first tentative step in this direction has been taken by the compilation of a world mangrove atlas (Spalding et al., 1997). This exercize demonstrated that there is a paucity of reliable data and that more effort is required at the level of individual countries to catalogue the extent of the mangrove resource.

Ground based techniques are of limited use in mapping mangroves because movement through such areas is difficult. Remote sensing techniques, using aircraft or satellites, avoid these problems. It is possible to measure the spatial extent of different mangrove habitats and canopy characteristics such as leaf area index. However, the cost of applying these techniques at a global scale is prohibitive.

Remote sensing is useful in the case of mapping defined regions or detailed habitats, though ground truthing is still required. The transfer of such data to a GIS enables information to be updated rapidly and derived from multiple sources. Subsequently, comparative analytical work can be undertaken.

The use of remote sensing and GIS in rehabilitation projects is practically non-existent. This is probably because the scale of rehabilitation projects is so small. However, GIS could play an important role in monitoring rehabilitation projects and will probably become more important if mangrove rehabilitation becomes more extensive.

Remote sensing and GIS involving mangrove ecosystems have important roles to play in quantifying the extent, structure and development of mangrove ecosystems and, though it is a relatively expensive activity, it deserves more attention.

Human factors and mangrove rehabilitation

Human factors critically influence the outcomes of mangrove rehabilitation projects in populated areas. It is recognised that human involvement in causing degradation is as important as the natural determinants of the structure and function of the mangrove ecosystem in determining the success of a rehabilitation project.

It follows that for a mangrove rehabilitation project to be effective, the human population must be viewed as an inherent part of the mangrove ecosystem. This means that the act of rehabilitation must benefit the indigenous people that depend on the exploitation of the ecosystem, use the relevant local knowledge and skills and effectively employ the existing social and political structures.

Examples of concern for the human factor in mangrove rehabilitation projects are: the Philippines (Walters, 1997); Thailand (Wetlands International Asia-Pacific, 1997), India (M.S. Swaminathan Research Foundation, 1998) and Pakistan (M. Quereshi, 1998, pers. com.).

People working on large scale mangrove rehabilitation projects in poor, highly populated tropical environments consider the most important requirement is to factor in the involvement of the human community. Only by managing the rehabilitation project with the support and involvement of the local people and by using innovative approaches will the project be successful.

Databanks

There is clearly a need to develop some form of global mangrove data base and information system. Access to mangrove literature in refereed journals is relatively straight forward but getting access to the 'grey' literature in the form of reports from international agencies and consultants is much more problematic. For example, it is difficult to gauge how successful rehabilitation projects have been globally and what lessons have been learnt from failure. It is clearly impossible to carry out a critical review without access to the myriad of reports that remain in the archives of sponsoring agencies. It seems that many mangrove rehabilitation programmes are being carried out without any reference to lessons that might be learnt from other similar programmes. One must suspect a great duplication of effort. In addition there is probably very little external critical analysis of the worth of many of the projects and few of the results are ever published in refereed journals.

The design of a mangrove data base, or more likely data bases, presents some formidable problems. The nature and requirements of the potential user must be clearly identified. The data base must be reliable, current and, where appropriate, referenced. It must also be constantly monitored and easily accessed. It is unlikely that a single mangrove data base could be all things to all users. However, a selected data base, such as maps with accompanying information, could be of real use. It also seems important that the provision of scientific data is not confused with providing general information.

There is a real need for an archival system to be established where reports on mangrove rehabilitation can be lodged and accessed easily by interested people. The Internet may offer a partial solution. However, a more organised system needs to be created by one of the international agencies, in concert with other agencies, as considerable resources would be required for maintenance.

The International Society for Mangrove Ecosystems (ISME) is currently mounting a project for a Global Mangrove Database and Information System (GLOMIS) with a budget of over two million U.S. dollars for five years. It would seem appropriate that people with an interest in mangroves should make their views known to ISME.

Conclusions

The current situation with respect to the rehabilitation of mangrove ecosystems is that there is a substantial body of knowledge of silviculture, site selection and monitoring. This should be built upon, not reworked. There is also a growing literature on the enhancement, rehabilitation and creation of coastal wetlands (Turner & Lewis, 1997), including mangroves, which needs to be extended.

It is vital to stress the importance of identifying the objectives of carrying out a rehabilitation programme and to integrate such objectives with the culture and welfare of the local communities dependent on the mangrove ecosystem for sustenance. There is a need for mangrove research and development activities to focus on areas that will support and enhance mangrove rehabilitation projects.

Finally, we may choose to rehabilitate mangrove ecosystems so that human populations can survive or we may choose to rehabilitate mangrove ecosystems to protect a rich natural inheritance for its own intrinsic value or we may choose to do both. However, as pressures from world over-population mount it is certain that strategies for the survival of human populations will become a necessity. Advocates of conservation for its own sake will face diminishing choices in a world of shrinking resources.

References

Brown, S. & A. E. Lugo, 1994. Rehabilitation of Tropical Lands: a key to sustaining development. Restoration Ecol. 2: 97–111.

Chan, H. T., 1996. Mangrove restoration in Peninsular Malaysia. In C. D. Field (ed.), Restoration of Mangrove Ecosystems. International Society for Mangrove Ecosystems, Okinawa, Japan: 64–75.

Chen, R. & R. R. Twilley, 1998. A gap dynamic model of mangrove forest development along gradients of soil salinity and nutrient resources. J. Ecol. 86: 37–51.

Davie, J. & R. Hynes, 1997. Integrating nature conservation and sustainable rural management. Aust. Biol. 10: 185–199.

Field, C. D. (ed.), 1996. Restoration of Mangrove Ecosystems. International Society for Mangrove Ecosystems, Okinawa, Japan: 250 pp.

JAM, 1997. Final report of the ITTO project on Development and Dissemination of Re-afforestation Techniques of Mangrove Forests [PD 11/92 Rev.1(F)]. Publication of the Japan Association for Mangroves, Tokyo, Japan: 104 pp.

Johnson, K., K. A. Vogt, H. J. Clark, O. J. Schmitz & D. J. Vogt, 1996. Biodiversity and the productivity and stability of ecosystems. Trends Ecol. Evolut. 11: 372–377.

Lakshmi, M., S. Rajalakshmi, M. Paranii, C. S. Anuratha & A. Parida, 1997. Molecular phylogeny of mangroves 1. Use of molecular markers I assessing the intraspecific genetic variability in the mangrove species *Acanthus ilicifolius* Linn. (Acanthaceae). Theor. Appl. Genet. 94: 1121–1127

Lewis, R. R., 1994. Enhancement, Restoration and Creation of Coastal Wetlands. In Kent, D. M. (ed.), Applied Wetlands Science and Technology. . CRC Press: 167–191.

Robertson, A. I. & M. J. Phillips, 1995. Mangroves as filters of shrimp pond effluent: predictions and biogeochemical research needs. Hydrobiologia 295: 311–321.

Saenger, P. & N. A. Siddiqi, 1993. Land from the sea: the mangrove afforestation program of Bangladesh. Ocean Coast. Mgmt. 20: 23–39.

Spalding, M. D., F. Blasco & C. D. Field (eds), 1997. World Mangrove Atlas. International Society for Mangrove Ecosystems, Okinawa, Japan: 178 pp.

Swaminathan Research Foundation, 1998. Coastal Wetlands: Mangrove Conservation and Management. Annual Reports 1, 2, 3. M.S.Swaminathan Research Foundation. Madras, India.

Turner, R. E. & R. R. Lewis III, 1997. Hydrologic restoration of coastal wetlands. Special issue: Hydrologic Restoration of Coastal Wetlands. Wetlands Ecol. M. 4: 65–72.

Twilley, R. R., S. C. Snedaker, A. Yanez-Aranciba & E. Medina, 1996. Biodiversity and ecosystem processes in tropical estuaries: perspectives from mangrove ecosystems. In H. Money, H. Cushman & E. Medina (eds), Biodiversity and Ecosystem Functions: a Global Perspective. John Wiley and Sons, New York: 327–370.

Twilley, R. R., V. H. Rivera-Monroy, R. Chen & L. Botero, 1998. Adapting an ecological mangrove model to simulate trajectories in restoration ecology. Mar. Pollut. Bull. 37 (8–12): 404–419.

Underwood, A. J., 1995. Ecological research and (and research into) environmental management. Ecol. Appl. 5: 232–247.

Walters, B. B., 1997. Human ecological questions for tropical restoration: experiences from planting native upland trees and mangroves in the Philippines. Forest Ecol. Mgmt. 99: 275–290.

Wetlands International Asia-Pacific, 1997. Community Participation in Mangrove Forest Management and Rehabilitation in Southern Thailand. Second Interim Report (ECU funded project): 26 pp.

Hydrobiologia **413:** 53–62, 1999.
R.S. Dodd (ed.), Diversity and Function in Mangrove Ecosystems.

Phenological patterns of *Rhizophora* L., *Avicennia* L. and *Laguncularia* Gaertn. f. in Amazonian mangrove swamps

Marcus E. B. Fernandes
Department of Biology, University of York, York, YO1 5DD, U.K.
Current Address: Universidade Federal do Maranhão, Depto. de Biologia, Campus Universitário do Bacanga, Av. dos Portugueses S/N, São Luís-Maranhão, 65080-40 Brasil

Key words: phenology, *Rhizophora*, *Avicennia*, *Laguncularia*, Amazonia, Brazil

Abstract

The phenology of flowering and fruiting of *Rhizophora mangle* L., *R. harrisonii* Leechman, *Avicennia germinans* (L.) Stearn and *Laguncularia racemosa* (L.) Gaertn. f., were studied on Maracá Island, a continental island situated between Cape Orange (04° 20′ N) and Cape North (01° 40′ N) in the state of Amapá, North Brazil. The study was carried out during a complete annual cycle from May 1995 to April 1996. All the phenophases were produced throughout the entire year at almost all of the six studied sites. However, they presented significant peaks which characterized seasonality for all of the measured variables focused in this study. *Rhizophora mangle, Avicennia germinans* and *Laguncularia racemosa* showed significant annual variation in flowering. By contrast, *Rhizophora harrisonii* presented evenly distributed flowering throughout the same period of observation. The fruiting patterns showed significant variation for all sites. In general, species presented flowering peaks from August to January when temperatures were high. Fruiting peaked from October to April and occurred mostly during the wet period. Although fruiting has correlated significantly with the wet period, the correlation was less significant than that between flowering and the dry period. The flowering burst during the driest period seems to complement the strategy in which massive fruit production is obtained during the wet period. Consequently, it may ensure the development of propagules and the continued occupation of mangrove forests along the intertidal zone in the Amazon region.

Introduction

Information on phenological patterns has been obtained from intensive studies in a great variety of ecosystems throughout the world. The majority of these studies have correlated phenological events with biotic and abiotic factors in order to describe particular or general phenological trends (Frankie et al., 1974; Wielgolaski, 1974; French & Sauer, 1974; Malaisse, 1974; Monasterio & Sarmiento, 1976; Tomlinson, 1986; Boinski & Fowler, 1989; Seghieri et al., 1995; Morellato & Leitão-Filho, 1996; Brooke et al., 1996).

The phenology of mangrove forests has also been studied in many tropical regions (Gill & Tomlinson, 1971; Christensen, 1978; Wium-Andersen, 1981; Duke et al., 1984). They have mainly emphasized the relevance of the phenological patterns to stand production and the seasonal variation of these patterns in regard to the environmental conditions. Other studies have pointed out that the mangrove evergreen and ever-growing patterns are related to variations in groundwater salinity (Christensen & Wium-Andersen, 1977; Wium-Andersen & Christensen, 1978). Moreover, it is well known that availability of water is a limiting factor which forces each species to optimize its use of water in time and space (Seghieri et al., 1995). In fact, the physical environment plays an important role in triggering phenological patterns in tropical forests (Liebermann, 1982). Consequently, the phenological rhythms tend to be flexible and, in turn, phenophases develop according to strategies which are synchronized with the local or regional environmental conditions that invariably regulate the system.

In spite of the relevance of the phenological events in the mangrove system, due to their intrinsic relationship with litter production and reproductive success of the species, few authors have dealt with the phenological aspects of this forest. As far as the Amazonian mangrove forests are concerned, no previous studies have been carried out to understand the main factors that are associated with the high productivity and the establishment of these forests on the intertidal zones along the coastline. This study is an effort to fill this gap and hence better understand the functional aspects inherent to this particular region. The part of the Amazon coastline where this study was carried out has a hot and humid climate. The mean annual precipitation is about 3250 mm and the annual potential evapotranspiration is about 1300 mm (Schaeffer-Novelli & Cintron, 1988). Low rates of precipitation are also recorded but without a definite drought, i.e. the values for precipitation on a monthly basis are always higher than 100 mm. The high shoreline avoids constant inundation by the tides, with exception of the spring tides during some months in the wet season. By studying the phenology of *Rhizophora mangle* L., *R. harrisonii* Leechman, *Avicennia germinans* (L.) Stearn and *Laguncularia racemosa* (L.) Gaertn. f., along an entire annual cycle, this study aimed to determine the periodicity as well as the environmental conditions leading to flowering and fruiting. In general, all the mangrove species focused on in this study showed a conspicuous seasonality regarding phenophase peaks. These findings are discussed in the light of the general trends that regulate reproductive production in the mangrove system.

Materials and methods

Phenological patterns

The phenological study recorded data from *Rhizophora mangle, Rhizophora harrisonii, Avicennia germinans* and *Laguncularia racemosa* as the focal species in the Ecological Station of Maracá-Jipioca Islands (Figure 1). They are the four main mangrove species which dominate the landscape on the coastline of the state of Amapá, North Brazil.

Phenological observations were recorded on selected trees on the six sites previously established for the whole study. The state of the reproductive organs (flowers and fruits) was recorded. Fifty mature individual specimens on each site were tagged and observed on a monthly basis from May 1995 to April 1996. Observations on *Rhizophora* spp. and *Avicennia* sp. were carried out by using binoculars due to the trees height (which measured on average 25 and 30 m, respectively), with exception of *Laguncularia* sp. which measured about 8 m. By the end of the study, a total of 12 observations per plant represented a complete annual cycle. Considering that each site is dominated by one of the mangrove species, the number of individual specimens recorded per species was: *Rhizophora mangle* (N=50 – Site#01); *Rhizophora harrisonii* (N=50 – Site#02); *Avicennia germinans* (N=100 – Site#03 and Site#04); and *Laguncularia racemosa* (N=100 – Site#05 and Site#06), totalling 300 trees among the study sites and 3600 observations throughout the year.

Phenophase observation

Given the structural characteristics presented by the mangrove stands on Maracá Island, i.e. due to the heights of the branches above the ground and high canopies, the assessment of flowering and fruiting was carried out according to the technique described by the National Research Council-U.S.(1981). Each phenophase is estimated as a percentage of the canopy. It means that percentages are estimated broadly on an ordinal scale. At each visit (N=12 during the whole study), the numbers of flowers and fruits on each marked tree specimen were assessed according to the following ordinal scale:

0= no flowers or fruits
1= from 1 to 10 item (s)
2= from 11 to 50 items
3= from 51 to 100 items
4= more than 100

Climatic data

Meteorological data were provided by the Brazilian Institute of Meteorology for the entire year (May 1995 – April 1996) in which the fieldwork was carried out. Precipitation varied throughout the year but was always high values (over 100 mm per month), hence there is no drought in this region. However, a relatively 'dry period' occurs from August to October, where lower rates of precipitation are recorded. Temperature fluctuated (around 1–2 °C) from month to month but did not present any obvious pattern throughout the year (Figure 2). Four abiotic variables (insolation, mean temperature, precipitation and speed of the wind) were used in order to investigate whether the appearance of a given phenophase was related to some environmental parameter.

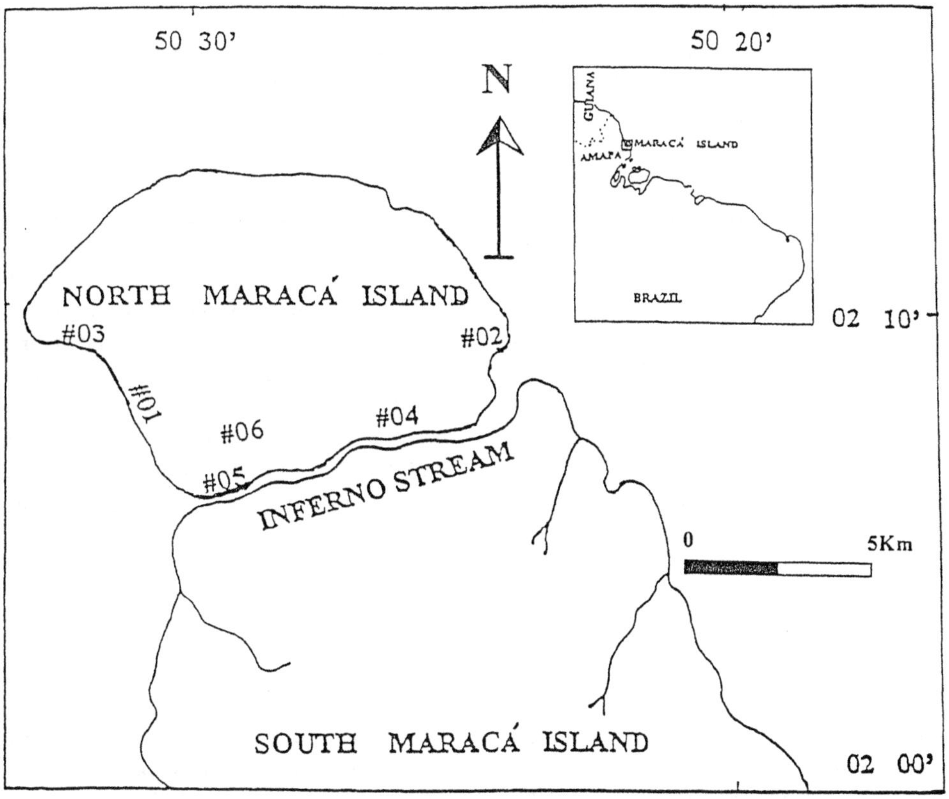

Figure 1. Map of the Maracá Island (Amapá, Brazil) showing the locations of the six mangrove studied sites.

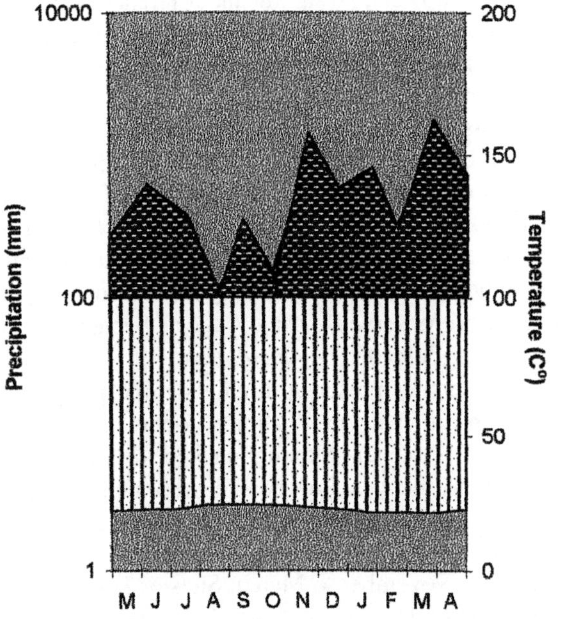

Figure 2. Climatic diagram from Maracá Island, May 1995 – April 1996. It shows mean temperature (lower) and total precipitation records (upper shaded area). Area with vertical bars below 100 mm precipitation means humid period, above, high precipitation.

Statistical analysis

To assess seasonality in flowering or fruiting, the values of the ordinal scale between zero and four were considered for non-parametric statistics (Siegel, 1975; Gibbons, 1976; Campbell, 1989). The value of each tree per month, during a complete year in each site, was treated by Friedman two-way ANOVA (K related samples), corrected for ties. Data for a comparative analysis between sites from the same botanical genus were treated by Kolmogorov-Smirnov Test (two independent samples). For the comparison among the three mangrove genera, data from the sites which presented dominance of the same botanical genus were pooled together and assessed by Kruskal-Wallis Test one-way ANOVA (K independent samples). Significance levels were determined after correcting for ties. The relationship between the abiotic factors (insolation, mean temperature, precipitation and wind speed) and the appearance of phenophases (flowering/fruiting) was assessed firstly by Stepwise Regression (Monk, 1991). However, multicollinearity was detected by high values of the Condition Index (CI) in the regression model (Belsley et al., 1980). Therefore, Principal Components Analysis (PCA) was used to avoid high multicollinear variables in the model. The PCA produced one principal component (factor) which was used as an index (hereafter Environmental Factor), in which all the environmental variables are weighted. Consequently, the Environmental Factor does not present multicollinearity and a Linear Regression was performed using the Environmental Factor as a dependent variable of the model. This technique allowed to assess whether environmental variables presented significant relationship to flowering or fruiting seasonality to each site. The most significant environmental variable to influence flowering or fruiting was determined according to the highest Pearson Correlation value presented by the former Stepwise Regression.

Results

Seasonality of flowering

The three mangrove genera focused on in this study (*Rhizophora, Avicennia* and *Laguncularia*), yielded flowers throughout the entire annual cycle, with the exception of *Avicennia*. They presented a highly significant difference in monthly total production when compared to each other (Kruskal-Wallis Test, χ^2 = 33.17, 2 df, $p<0.001$, Figure 3). However, such differences are not statistically significant when sites from the same dominant botanical genus are compared, i.e. comparison between Site#01 and Site#02 – *Rhizophora*; Site#03 and Site#04 – *Avicennia*; Site#05 and Site#06 – *Laguncularia* (Kolmogorov-Smirnov Test, for all pairs of sites $p>0.05$).

The majority of sites showed significant seasonality in flowering. The exception was Site#02 – *Rhizophora harrisonii*, where almost all the tagged trees yielded a great amount of flowers throughout the year, with a massive production mainly during October and November (Friedman ANOVA, $p>0.05$, Table 1). A similar massive production of flowers was observed on Site#01 – *Rhizophora mangle* throughout the year. However, the trees presented significant variation of the flowers production in a monthly basis, therefore seasonality, in this case, was statistically high significant (Friedman ANOVA, $p<0.001$, Table 1).

Marked trees on Site#04 (*Avicennia germinans*) developed flowers during almost all of the months throughout the year. They also presented a visible synchrony with Site#03 (*Avicennia germinans*), on which none of the marked trees developed any flower during four months (May-1995, February, March and April-1996, Table 1). On Sites #05 and #06 (*Laguncularia racemosa*) no synchrony was observed. At the former site, flowers developed throughout the whole year (bearing in mind that during September and March only one tagged tree was observed flowering), whilst at the latter none of the marked trees flowered during August, September and April (Table 1). A phenological spectrum (Figure 4) shows the flower production throughout the year based on the mean of the 50 trees in each study site. All the plant species presented a peak of flowering about the same period, from August to January. It coincides mainly with the driest period, despite the fact that this region does not present any conspicuous drought (see Figure 2). This finding agrees with the fact that more plant species take advantage of the dry season for flowering (Croat, 1975).

Seasonality of fruiting

Fruits showed as high seasonal variation as flowers. The comparison among genera (*Rhizophora, Avicennia* and *laguncularia*), showed a significant variation in monthly total yield (Kruskal-Wallis Test, χ^2 =14.93, 2 df, $p < 0.01$, Figure 5). However, as observed for flowers, no significant difference was found when sites, with the same dominant botanical genus, were compared (Kolmogorov-Smirnov Test, for all pairs of sites $p>0.05$).

Fruits were produced almost throughout the whole year (Table 2). Sites of *Rhizophora*, for example, yield fruits throughout the year, whilst sites of *Avicennia* failed to develop fruits only in June (Site#04 – Figure 4). Site#05 (*Laguncularia*) also produced fruits throughout the entire annual cycle. In contrast, no fruits were observed during August, September and October on Site#06 (*Laguncularia*).

All the sites presented highly significant monthly variation of fruit production (Friedman ANOVA, $p<0.001$, Table 2). Although tagged trees developed fruit along the year, every site presented a defined peak of fruiting. Sites of *Rhizophora* showed a peak from October to April, whilst on *Avicennia* sites it was from November to March. Sites of *Laguncularia* peaked from December to April. In general, fruiting and flowering events were developed throughout the annual cycle. Furthermore, the results indicate that these events for the three mangrove genera overlap along the year. However, there is a visible tendency

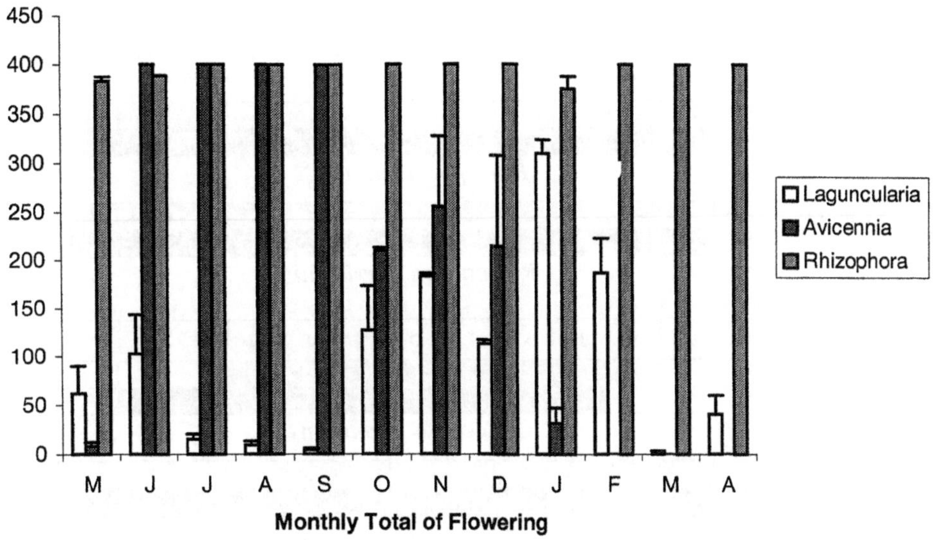

Figure 3. Summary of flowering observations of *Rhizophora, Avicennia* and *Laguncularia* on Maracá Island. The graph shows the monthly total of flowering for each mangrove genus observed. The values of the scale from 0 to 4 (see 'Materials and methods') were summed, thus the maximum value to be reached is 400 (50 trees per site; 2 sites for each botanical genus).

Table 1. Seasonality of Flowering of *Rhizophora mangle, R. harrisonii, Avicennia germinans* and *Laguncularia racemosa* in the six sites on Maracá Island. It shows the number of tagged plants (n=50 in each site), which produced flowers in the month. It also shows the months of greatest flower abundance, together with the χ^2 value of each site, with the significance of the Friedman ANOVA, according to the production of the total tagged individuals on a monthly basis. *= p<0.05, **= p<0.01, ***= p<0.001, NS= Not Significant

Site (species)	Tagged plants flowering in month												Months of greatest	χ^2 Value and
	M	J	J	A	S	O	N	D	J	F	M	A	Flower abundance	signif. (11 df)
#01 (*Rhizophora mangle*)	48	49	50	50	50	50	50	50	50	50	50	50	Oct, Dec	19.9 ***
#02 (*Rhizophora harrisonii*)	49	49	50	50	50	50	50	50	50	50	50	50	Oct, Nov	1.7 / NS
#03 (*Avicennia germinans*)	0	50	50	50	50	50	36	14	30	0	0	0	Aug, Sep, Oct	51.7 ***
#04 (*Avicennia germinans*)	2	50	50	50	50	50	50	50	0	0	0	0	Sep, Oct, Nov	54.4 ***
#05 (*Laguncularia racemosa*)	19	36	2	4	1	39	28	24	39	39	1	18	Oct, Jan	22.7 ***
#06 (*Laguncularia racemosa*)	2	4	6	0	0	10	34	26	48	26	2	0	Nov, Jan	31.6 ***

in which, mainly during the transitory period from dry to wet season, the peak of flowering coincides mostly with the dry period whilst fruiting extend through the wet season (Figure 4).

Climatic data

The peak of flowering occurred between the months of August and January which partially correspond with the lowest intensity of rainfall and highest records for mean temperature (Figure 2). Figure 6 shows the relationship of the abiotic variables with flowering at the study sites on Maracá Island. Site#03 (*Avicennia germinans*) presented highly significant relationship with the mean temperature ($F(1,10)=36.8$, $p<0.001$). On the other hand, Site#01 (*Rhizophora mangle*) was significantly related to insolation ($F(1,10)=15.3$, $p<0.01$). In fact, flowering on these sites was positively related to the dry period in which insolation seems to take an important role.

There was no evident tendency of fruits to peak in any particular season. At Site#01 – *Rhizophora mangle* – for example, it was negatively related to precipitation ($F(1,10)=6.0$, $p<0.05$), whilst Site#06 – *Laguncularia racemosa* – presented a negative relationship with insolation ($F(1,10)=5.2$, $p<0.05$, Figure 7).

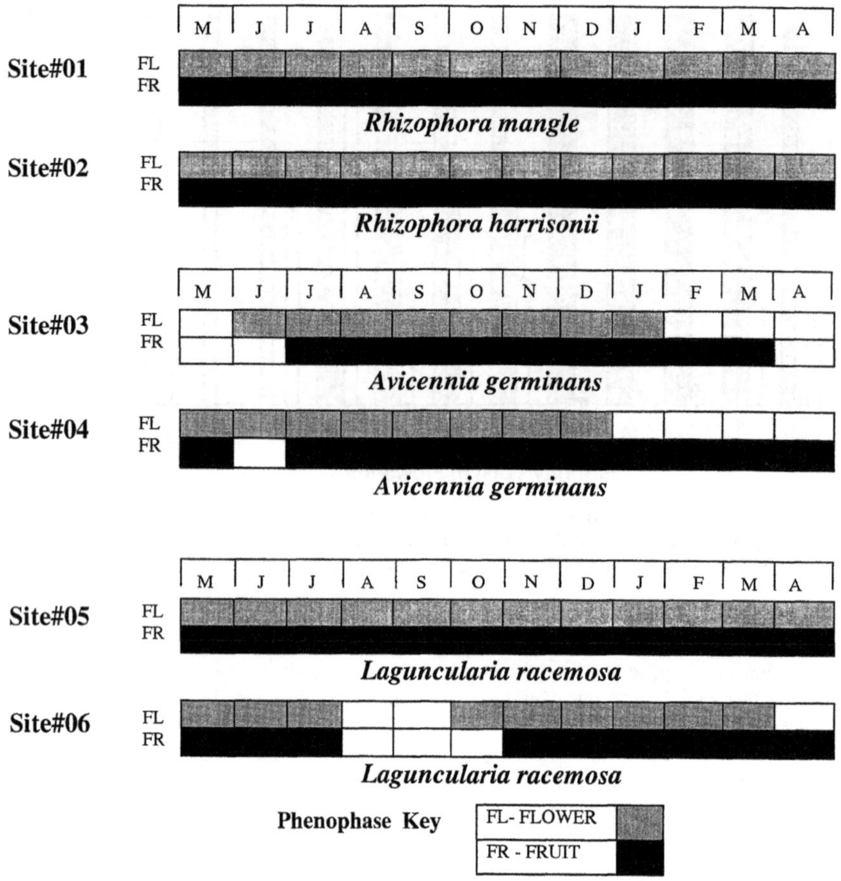

Figure 4. Phenological spectrum for the mangrove genera *Rhizophora, Avicennia* and *Laguncularia* in the six sites on Maracá Island. Phenophases are represented in rows as Flower (FL) and Fruit (FR – unripe and/or ripe) for each site. Clear and shaded boxes represent the absence and presence, respectively, of the phenophase on a monthly basis.

Table 2. Seasonality of Fruiting of *Rhizophora mangle, R. harrisonii, Avicennia germinans* and *Laguncularia racemosa* in the six sites on Maracá Island. It shows the number of tagged plants (n=50 in each site), which produced flowers in the month. It also shows the months of greatest flower abundance, together with the χ^2 value of each site, with the significance of the Friedman ANOVA, according to the production of the total tagged individuals on a monthly basis. * = $p < 0.05$, **= $p<0.01$, ***= $p<0.001$, NS= Not Significant

Site (species)	Tagged plants fruiting in month M	J	J	A	S	O	N	D	J	F	M	A	Months of greatest fruit abundance	χ^2 Value and signif. (11 df)
#01 (*Rhizophora mangle*)	50	43	50	50	50	50	50	50	50	50	50	50	Jan, Feb	46.6 ***
#02 (*Rhizophora harrisonii*)	16	32	50	47	50	50	50	50	50	50	50	50	Oct, Apr	48.7 ***
#03 (*Avicennia germinans*)	0	0	50	50	50	50	50	50	50	40	17	0	Dec, Jan	53.1 ***
#04 (*Avicennia germinans*)	50	0	50	50	50	50	50	50	50	50	50	50	Nov, Mar	54.9 ***
#05 (*Laguncularia racemosa*)	48	48	41	4	5	11	39	43	44	48	50	48	Dec, Jan	25.7 ***
#06 (*Laguncularia racemosa*)	21	4	4	0	0	0	9	32	11	26	50	50	Mar, Apr	46.1 ***

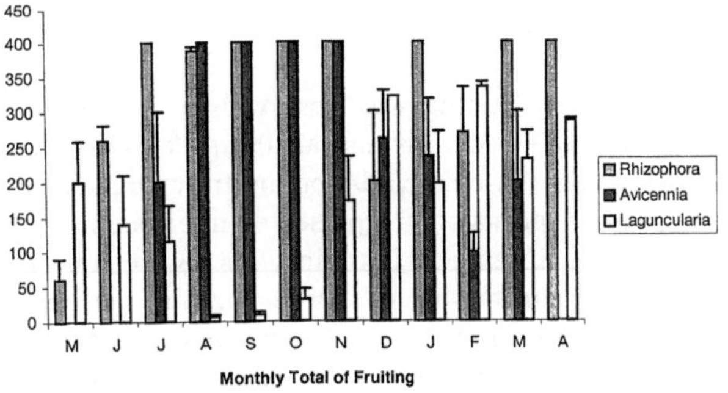

Figure 5. Summary of fruiting observations of *Rhizophora, Avicennia* and *Laguncularia* on Maracá Island. The graph shows the monthly total of flowering for each mangrove genus observed. The values of the scale from 0 to 4 (see 'Materials and methods') were summed, thus the maximum value to be reached is 400 (50 trees per site; 2 sites for each botanical genus).

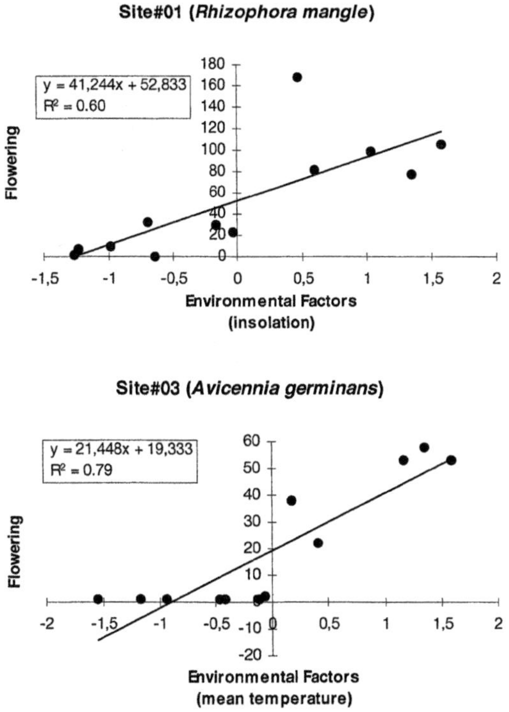

Figure 6. Relationship between the flowering period and the environmental parameters for the sites which showed significant *p* values by linear regression. Between parenthesis is represented the most significant environmental variable according to Pearson Correlation.

Discussion

Reproductive phenology of the four main mangrove species (*Rhizophora mangle, R. harrisonii, Avicennia germinans* and *Laguncularia racemosa)* was studied along an entire year on Maracá Island. The geographical situation of the island, close to the Equator line, is characterized by a constant solar radiation and day length, slight variation in temperature and irregular fluctuation of the rainfall throughout the annual cycle. Hence, it is reasonable to expect that such abiotic factors might play an important role in the phenological patterns presented by these mangrove plant species. However, the factors which promote such phenological events leaves open the question of whether these patterns are related to abiotic or biotic factors, or even a complex of both.

Seasonality of flowering

Although the majority of the mangrove species has produced all the items (phenophases) through the year, the results indicated that some fluctuations occurred on a monthly basis, indicating a seasonal cycle with visible peaks (Tables 1).

Avicennia was the first to present a visible peak along the year followed by *Rhizophora* and *Laguncularia*. This phenological pattern was characterized by a concomitant flowering peak at all sites which coincided with the months of highest temperature. Despite Site#01(*R.mangle*) and Site#03(*A.germinans*) showing significant correlations with environmental factors (Figure 6), the other four sites studied, including Site#04 which is also dominated by *A.germinans*, did not show any relationship with the abiotic events. It appears that environmental conditions exert some influence on flower production of some species, mainly in *R.mangle* and *A.germinans*. In contrast, *L.racemosa, R. harrisonii* and even *A.germinans* on Site#04 did not present any correlation with those environmental variables, indicating that other factors promote flowering on the island.

However it is reasonable to point out that flowering followed a general trend presenting peaks which coincided with the dry period. This is in agreement with other mangrove species studied in different areas (Saenger, 1982; Steinke & Rajh, 1995). In western Africa, for example, in spite of the lack of systematic data on phenology of mangrove species, data compiled by Saenger & Bellan (1995), showed that the principal mangrove species (*Rhizophora mangle, R.racemosa, R. harrisonii, Avicennia germinans* and *Laguncularia racemosa*) appear to yield flowers during the dry period of the year, following this general trend.

It is well known that mangroves are adapted to a generalised type of pollinators and thus competition for the available pollinators is reduced (Tomlinson,

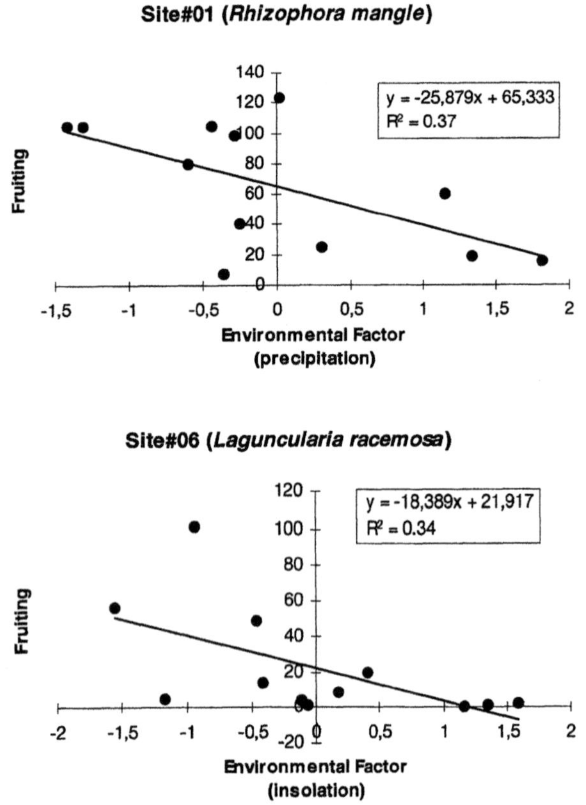

Figure 7. Relationship between the fruiting period and the environmental parameters for the sites which showed significant p values by linear regression. Between parenthesis is represented the most significant environmental variable according to Pearson Correlation.

1986). Furthermore, this extended flowering period is a strategy that may be considered as an important adaptation to the foraging behaviour displayed by the pollinators (Janzen, 1971). Since flowers are available through the year, in turn, pollinators are able to perform foraging activities on a regular basis. Considering that a widely spread series of plants of the same species can be visited by a large spectrum of potential pollinators, mainly insects, competition tends to be minimised. Mangrove stands are often dominated by few plant species or even present a monospecific structure and the frequent movements of the pollinators allow for effective outcrossing of these species. In addition, visitation of pollinators could be considered a constraint for fruit-flower ratio, since flowers which receive less pollen are more likely to fall, rather than the others with more pollen which develop into fruits (Murali, 1993).

Seasonality of fruiting

As flowering, fruiting was observed in all sites almost in a monthly basis and all the species showed a significant seasonal variation in fruit abundance. The seasonal peaks of fruit presented interspecific and intraspecific differences in timing. They were concentrated between the end of dry period and the onset of the wet season (Table 2). There was no synchrony between sites of the same species, ex. *A. germinans*, which peaked on December and on January at Site#03, whereas on Site#04 fruits were abundant in two different periods, November (end of the dry season) and March (rainy season). The same asynchrony was registered for the two *Rhizophora* species on Sites#01 (*R.mangle*) and Site#02 (*R. harrisonii*). In the case of *L.racemosa*, fruiting peaks coincided with the general trend in which fruiting tends to correlate with the rainy season (Saenger & Bellan, 1995). However, there was no significant positive correlation between fruiting and rainfall in this study. In contrast, a negative correlation was found between precipitation and fruiting on Site#01 (Figure 7). These results arose from the fact that the fruiting peaks were less steadily along the year and they were unlikely to reject the null hypothesis. However, low significant correlation was found between Site#04 and precipitation. In fact, it suggests a slight tendency for some species to fruit in the wet period.

Since mangrove species have water-dispersed propagules, it is reasonable to expect fruiting peaks to occur mostly during the rainy season on Maracá Island. During the rainy period, all the six sites become flooded by the rain water forming temporary ponds from where litter (e.g. leaves, twigs, flowers, propagules, stipules, etc.) are flushed to adjacent areas, promoting dispersion of the propagules. Site#03 (*L. racemosa*), for example, must be considered an exception amongst the sites because seed dispersion by the tide movements occurs more often due to this site be placed on the lower shore. The fruits that were observed dropping along the dry season are much more likely to be predated by crabs and rodents (e.g. rats and agoutis) rather than developing into trees. Agoutis, for example, are effective predators of mangrove propagules but can only be seen in the stands during the dry period, whilst during the wet season they are restricted to the dry spots of 'tesos' (type of forest dominated by non-mangrove species) spread on the island. In addition, loss of viability of fruits in tropical areas tends to be accelerated by sunlight and, in contact with the soil, through attack by soil organisms (Kigomo et al., 1994).

Wium-Andersen & Christensen (1978) suggested that the phenological patterns of some mangrove species are influenced by salinity variations in the groundwater. In this study, the results showed that there are no significant differences in the phenological patterns between the two sites of *L.racemosa* (Kolmogorov -Smirnov Test, for flowering and fruiting $p > 0.05$). Both sites of *L.racemosa* (Site#05 and Site#06) present extremely different conditions regarding salinity. Site#05, for example, is placed seaward in the lower shore under the tidal influence, whilst Site#06 is situated landward without any tidal influence. In addition, the latter site becomes a 60 cm deep freshwater pond during the wet season. The same, not significant, difference was found for *Rhizophora* spp. and *Avicennia* sp. Consequently, salinity cannot be viewed as an important factor affecting phenological patterns on Maracá Island.

Rhizophora mangle, R. harrisonii, Avicennia germinans and *Laguncularia racemosa* presented dispersed pattern of flowering and fruiting throughout the year. And also considering that on the island there is no accentuated drought period (Figure 2), both facts tend to indicate that environmental factors, i.e. climate, water availability, etc. does not entirely by themselves influence the phenological events registered along the entire annual cycle. On the other hand, along with the abiotic factors, physiological adaptive features of these species might be a relevant factor that allow them to assure reproduction and the renewed occupation of niches along the Amazon shoreline.

Acknowledgements

I am grateful to Dr R. Ormond and Dr P. Hogarth of the University of York, U.K. for their useful comments on the draft of the manuscript. Sponsorship was from the National Research Council - CNPq, Brazil.

References

Belsley, D. A. E. Kuh & R. E. Welsch, 1980. Regression diagnostics: identifying ingluential data and source of collinearity, John Wiley & Sons, New York, Chap. 3.

Boinski, S. & N. L. Fowler, 1989. Seasonal patterns in a tropical lowland forest. Biotropica 21: 223–233.

Brooke, L. M., J. P. Jones, J. A. Vickery & S. Waldren, 1996. Seasonal patterns of leaf growth and loss, flowering and fruiting on a subtropical central pacific island. Biotropica 28: 164–179.

Campbell, R. C., 1989. Statistics for biologists. Cambridge University Press. 3rd Edn: 446 pp.

Christensen, B., 1978. Biomass and primary production of *Rhizophora apiculata* Bl. in a mangrove in southern Thailand. Aquat. Bot. 4: 43–52.

Christensen, B. & S. Wium-Andersen, 1977. Seasonal growth in mangrove trees in southern Thailand. I. The phenology of *Rhizophora apiculata* Bl. Aquat. Bot. 3: 281–286.

Croat, T. B., 1975. Phenological behavior of habit and habitat classes on Barro Colorado Island (Panama Canal Zone). Biotropica 7: 270–277.

Duke, N. C., J. S. Bunt & W. T. Williams, 1984. Observations on the floral and vegetative phenologies of North-eastern Australian mangroves. Aust. J. Bot. 32: 87–99.

Frankie, G. W., H. G. Baker & P. A. Opler, 1974. Comparative phenological studies of trees in tropical wet and dry forests in the lowlands of Costa Rica. J. Ecol. 62: 881–913.

French, N. & R. H. Sauer, 1974. Phenological studies and modeling in grasslands. In Lieth, H. (ed.), Phenology and Seasonality Modeling. Chapman & Hall, London: 227–236.

Gibbons, J. D., 1976. Nonparametric methods for quantitative analysis. Holt, Rinehart and Wisnton. 463 pp.

Gill, A. M. & P. B. Tomlinson, 1971. Studies of the growth of red mangrove (*Rhizophora mangle* L.). 3. Phenology of the shoot. Biotropica 3: 109–124.

Janzen, D. H., 1971. Euglossine bees as long-distance pollinators of tropical plants. Science, N.Y. 171: 203–205.

Kigomo, B. N., S. R. Woodell & P. S. Savill, 1994. Phenological patterns and some aspects of reproductive biology of *Brachylaena huillensis* O. Hoffm. Afr. J. Ecol. 32: 296–307.

Liebermann, D., 1982. Seasonality and phenology in a dry tropical forest in Ghana. J. Ecol. 70: 791–806.

Malaisse, F. P., 1974. Phenology of the Zambezian woodland area with emphasis on the miombo ecosystem. In Lieth, H. (ed.), Phenology and Seasonality Modeling. Chapman & Hall, London: 269–286.

Monasterio, M. & G. Sarmiento, 1976. Phenological strategies of plant species in the tropical savanna and the semi-deciduous forest of the Venezuelan llanos. J. Biogeogr 3: 325–356.

Monk, A., 1991. Exploring statistics with Minitab, a workbook for the behavioural sciences. Wiley & Sons: 247 pp.

Morellato, P. C. & H. F. Leitão-Filho, 1996. Reproductive phenology of climbers in a southeastern Brazilian forest. Biotropica 28: 180–191.

Murali, K. S., 1993. Differential reproductive success in *Cassia fistula* in different habitats – a case of pollinator limitation? Curr. Sci. 65: 270–272.

National Research Council – U.S., 1981. Techiniques for the study of primate population ecology. National Academy Press, Washington: 233 pp.

Saenger, P., 1982. Morphological, anatomical and reproductive adaptations of Australian mangroves. In Clough, B.F. (ed.), Mangrove ecosystems in Australia. ANU Press, Canberra.

Saenger, P. & M. F. Bellan, 1995. The mangrove vegetation of the atlantic coast of Africa: a review. Unpublished Report: 64 pp. (English Version).

Schaeffer-Novelli, Y. & G. M. Cintron, 1988. Expedicão nacional aos manguezais do Amapá, Ilha de Maracá. Unpublished Report: 106 pp.

Seghieri, J. Ch. Floret & R. Pontanier, 1995. Plant phenology in relation to water availability: herbaceous and woody species in the savannas of northern Cameroon. J. Trop. Ecol. 11: 237–254.

Siegel, S., 1975. Estatística não-paramétrica, McGraw - Hill, Editora Santuário, Brasil: 350 pp.

Steinke, T. D. & A. Rajh, 1995. Vegetative and floral phenology of the mangrove, *Ceriops tagal,* with observations on the reproductive behaviour of *Lumnitzera racemosa*, in the mgeni estuary. S.-Afr. Tydskr. Plantk. 61: 240–244.

Tomlinson, P. B., 1986. The botany of mangroves. Cambridge University Press. 413 pp.

Wielgolaski, F. E., 1974. Phenological studies in tundra. In Lieth, H. (ed.), Phenology and Seasonality Modeling. Chapman & Hall, London: 209–214. pp.

Wium-Andersen, S., 1981. Seasonal growth of mangrove trees in southern Thailand. III. Phenology of *Rhizophora mucronata* Lamk. and *Scyphiphora hydrophyllacea* Gaertn. Aquat. Bot. 10: 371–376.

Wium-Andersen, S. & B. Christensen, 1978. Seasonal growth of mangrove trees in southern Thailand. II. Phenology of *Bruguiera cylindrica, Ceriops tagal, Lumnitzera littorea* and *Avicennia marina*. Aquat. Bot. 5: 383–390.

Hydrobiologia **413:** 63–76, 1999.
R.S. Dodd (ed.), Diversity and Function in Mangrove Ecosystems.

Red mangrove (*Rhizophora mangle* L.) litter fall response to selective pruning (Indian River Lagoon, Florida, U.S.A.)

Randall W. Parkinson[1], Monica Perez-Bedmar[1,2] & Jenna A. Santangelo[1]
[1]*Division of Marine and Environmental Systems, Florida Institute of Technology, 150 West University Blvd., Melbourne, FL 32901, U.S.A. Tel. [+1] 407-674 7455*
[2]*Current address: 710 Executive Center Drive, 1-34, West Palm Beach, FL 33401, U.S.A.*

Key words: above-ground productivity, estuarine food web, organic detritus, mechanical alteration

Abstract

This 33 month study quantified red mangrove (*Rhizophora mangle* L.) litter fall response to a selective pruning event using fringing forests located along the Indian River Lagoon, Florida, U.S.A. Selective pruning consisted of the removal of as many as 50% of the lateral branches originating between 2.1m (7ft) and 4.5m (15ft) above the forest floor while maintaining at least 50% of the forest canopy. Subcanopy light transmission data were used to estimate the impact of pruning on canopy closure and monthly measurements were obtained thereafter to monitor recovery. Estimates of litter fall were based upon biweekly sampling from ten 0.25 m^2 traps randomly placed in four 5 $\times$ 10 m control and impact plots. Following selective pruning, subcanopy light transmission increased by more than 30%. This provided a favorable environment for enhanced mangrove propagule recruitment, but several exotic species, including Brazilian Pepper (*Schinus terebinthifolius*), also invaded the forest beneath canopy gaps. Subcanopy light transmission within the impact plots has steadily declined since pruning and within 12 months had approached control plot levels. The results of our BACI impact assessment suggest *mean* forest litter fall production before and after pruning was not significantly different. However, inspection of the graphical output suggest all components of litter fall declined immediately after impact. Convergence towards pre-impact production levels took ~6 months. Many mangrove forests in peninsular Florida are subjected to multiple pruning events or other forms of repeated mechanical alteration by homeowners who seek to maintain a scenic vista of the Indian River Lagoon or other coastal waterway. The cumulative effect of stress induced by repeated impact is unknown at present.

Introduction

Over the past century, an estimated 150 000 acres of mangrove forest were eliminated during the rapid urbanization of peninsular Florida (Lewis et al., 1985). As data documenting their habitat value and ecosystem function became available (cf. Odum & Heald, 1972; Lugo & Snedaker, 1974; Cintron, 1982; Tomlinson, 1986), legislative and legal debates emerged over the economic and ecological consequence of widespread and unregulated forest destruction (Odum et al., 1985). In 1985, the Florida legislature responded to these concerns by establishing the 'mangrove protection rule'. This rule (ss403.93) defined when and how mechanical alteration (i.e. selective pruning) could be conducted by homeowners, most of whom were seeking a scenic vista of an adjacent coastal waterway.

The mangrove protection rule has undergone numerous amendments since inception, in part because existing data could not clearly delineate how regulated mechanical alteration was affecting the habitat value and ecosystem function of Florida's remaining mangrove forests. For example, some forms of mechanical alteration are beneficial to mangrove forest productivity. *Tree thinning*, a common silviculture technique, was known to reduce tree mortality and enhance seedling production when properly conducted (Putz & Chan, 1986; Teas, 1979; E. Heald, personal communication, 1994). *Selective pruning* can stimulate an increase in net primary productivity by inducing the development of new sprouts in most mangrove trees. The exception is the mangrove Rhizophoraceace, which loses the ability to produce reserve meristem within two or three years (Gill & Tomlinson, 1969, 1971) and therefore new sprouts will not develop from pruned branches >2.5 cm in diameter. This unique pruning response is one of the principle reasons regulated mechanical alteration of mangroves along shorelines of the Indian River Lagoon (IRL) has generated intense and persistent controversy. A vast majority of the intertidal vegetation within this estuary consists of red mangrove (*Rhizophora mangle* L.) and improper alteration could 1. reduce habitat value by altering forest structure or 2. disrupt ecosystem function by reducing leaf standing crop, litter fall production and ultimately the flux of organic matter to the detrital based estuarine food web (Jordan, 1971). The controversy was intensified by the IRL's successful nomination into the National Estuary Program and recognition as an Estuary of National Significance.

There have been a limited number of published reports assessing the effects of mechanical alteration on forest structure (Snedaker et al., 1992) and ecosystem function (Estevez & Evans, 1978; Beever, 1991). However, these analyses were conducted in mixed species forests located in peninsular Florida's southern and southwest coastal embayments. This study was designed to quantify litter fall response of red mangrove to selective pruning using fringing forests located on the IRL. Litter fall production was selected as a logical parameter to assess the impacts of selective pruning because it 1. has been shown to respond to stress induced by mechanical alteration, 2. is a measure of ecosystem function (i.e. primary productivity) and 3. is logistically and economically feasible to evaluate (Olsen, 1963; Cintron, 1982; Day et al., 1982, 1987; Proctor, 1983; Brown, 1984; Odum et al., 1985).

Experimental Design

Statistical method of impact assessment

To quantify the effects of selective pruning, litter fall data were evaluated using the **B**efore/**A**fter, **C**ontrol/**I**mpact (BACI) method described by Stewart-Oaten et al. (1986). The essence of the BACI approach is that the detection of an environmental impact is based on a test of the mean difference between control and impact groups compared before and after impact. In this analysis, the control (unpruned) and impact (pruned) groups were each composed of four 5 × 10 m experimental plots. Estimates of litter fall were obtained using 10 litter traps randomly placed in each plot. Collections occurred biweekly throughout a 33 month monitoring interval.

As an initial step, mean litter fall values obtained from the control and impact groups were graphed as a three-point moving average through time (Bunt, 1982). The graphical presentation of group means provided a simple method for revealing temporal trends (i.e. seasonal variation) or detecting erroneous data (i.e. outlier). Subsequently, the difference between control and impact group means was plotted as a function of time, also using a three-point moving average to facilitate visual evaluation of the graphical output. Finally, these data were used to test the following null hypothesis (H_0):

the difference between control and impact group means collected before selective pruning does not change significantly ($\alpha = 0.05$) after selective pruning

The H_0 test results, coupled with the descriptive statistics and graphical output, provided a reliable basis from which to assess the effects of selective pruning on litter fall production.

Site selection

Within the IRL, most mangrove forests subjected to mechanical alteration consist of monospecific stands of red mangrove (R. Mermer, personal communication, 1994). We inspected numerous sites before selecting a 20 m wide fringe forest (Lugo & Snedaker, 1974) of red mangrove located along the margin of St. Lucie County Mosquito Control Impoundment Number 2 (Figures 1 & 2). The forest consisted of a mature stand of undisturbed trees with a tall and dense canopy. An understory of perennial vegetation was absent. Successfully maturing seedlings were rare, occurring only beneath canopy caps. Several exotic species, especially Brazilian pepper (*Schinus terebinthifolius*), were present along the upland (dike) margin of the forest, but none of these species extend into the main forest area.

The initial criteria for selection of experimental plots within our study site included 1. similar forest structure, 2. accessibility, and 3. sufficient space to accommodate a 5 × 10 m monitoring area, an adjacent 15 m wide buffer zone (H. Swain, personal communic-

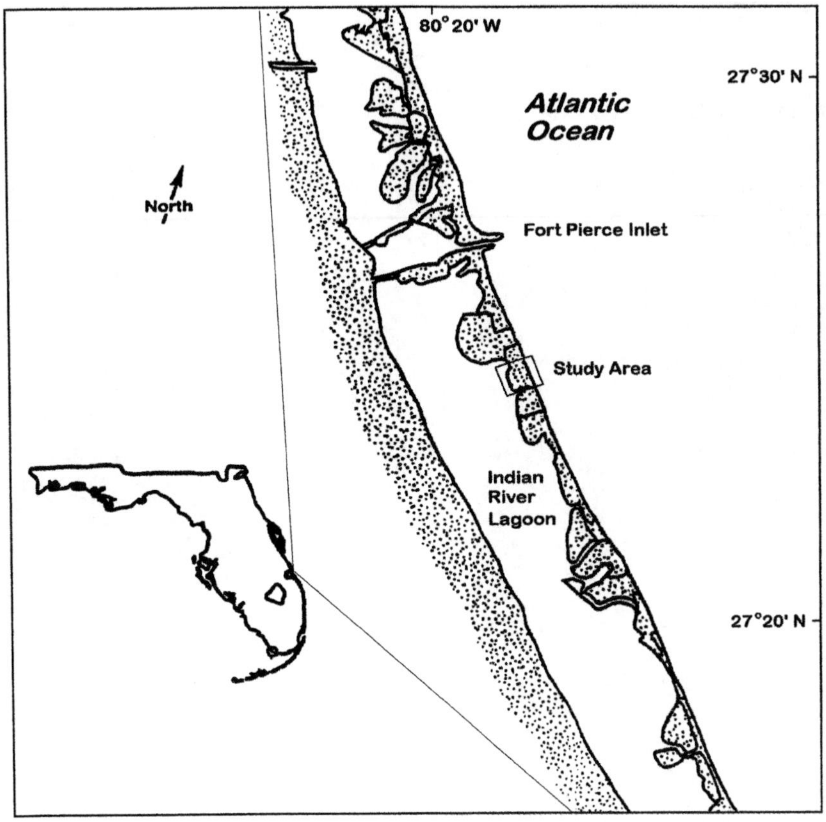

Figure 1. Regional map of study area, located on the Atlantic coast of peninsular Florida, a low relief carbonate platform. The wave-dominated, micro-tidal barrier island system provides extensive low-energy shorelines that are favorable to the establishment of tidal wetlands. The study area (box inset; see also Figure 2) is located at the northern limit of the subtropical climate zone, mangrove vegetation dominates the intertidal zone.

Table 1. Summary of *Rhizophora mangle* structural data obtained from control and impact plots at beginning of project (August 1993)

Group	N (DBH>2.5 cm)	DBH (cm)	Height (m)	Density (N/m^2)	Basal Area (m^2/ha)	Canopy Coverage (%)
Control	36	12.01	7.5	0.18	20.39	89
Impact	33	12.78	7.3	0.17	21.17	90

ation, 1995) and ten 0.25 m^2 litter baskets. Diameter breast height (DBH) data were used to identify suitable experimental plots because the parameter was easy to measure and representative of forest structure, primary productivity and litter fall production (cf. Day et al., 1982). A power test ($\alpha = 0.05$), conducted using DBH data (>2.5 cm) obtained from trees growing in two visually similar test plots, indicated control and impact groups each required a minimum of four experimental plots. Structural data (Table 1) were then collected until eight experimental plots were identified as acceptable subsamples of the forest population (Figure 2). Alternate plots ($N = 4$) were chosen for selective pruning to minimize geographical differences and prevent variations due to stochastic events (W. Nelson, pers. comm., 1994). The suitability of using four control and four impact plots at $\alpha = 0.05$ was later re-confirmed using litter fall data collected one and six months after the study had been initiated.

Selective pruning

Pruning of impact plots was conducted on 19 and 20 April 1995 following the general guidelines for *selective trimming* as defined by the Florida Department of Environmental Protection (DEP). As many as 50% of the lateral branches originating between 2.1 m (7 ft) and 4.5 m (15 ft) above the forest floor were selectively pruned from impact plots to create a *window* or scenic vista through which the adjacent coastal wa-

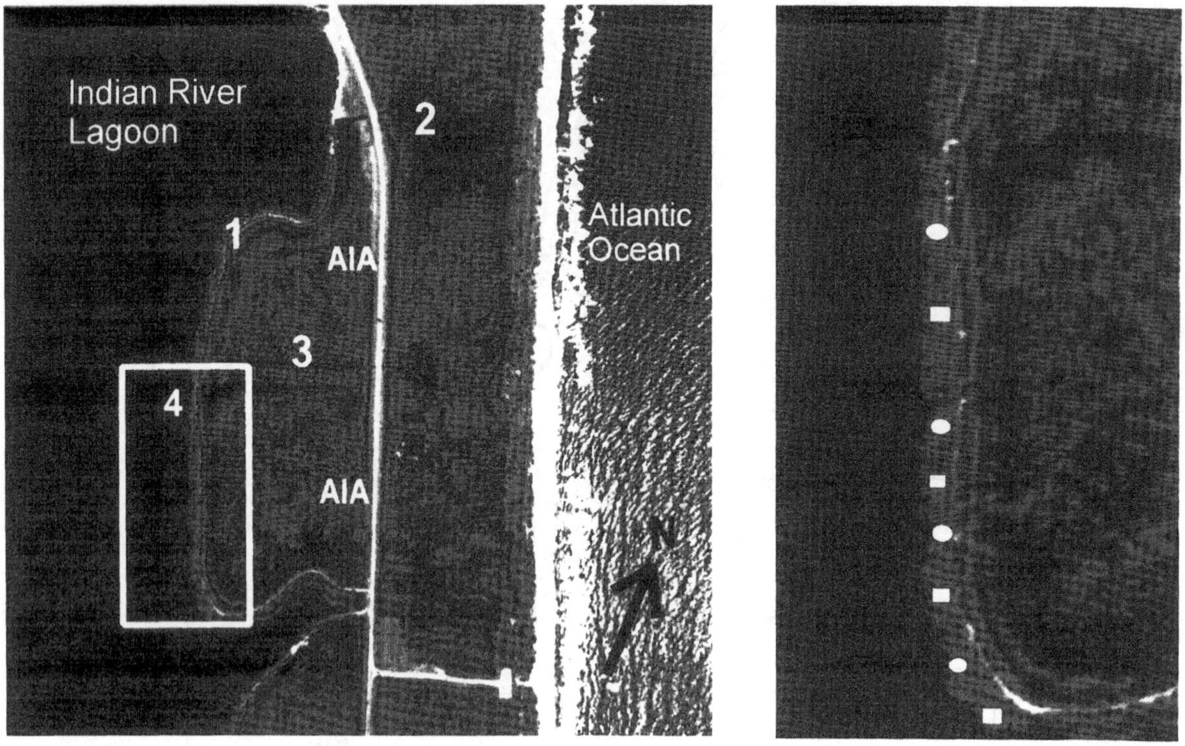

Figure 2. Black and white reproduction of infra-red image (1995) centered on study area (see Figure 1 for regional location). *Left:* St. Lucie County Mosquito Control Impoundment Number 2. Width of photo is 1.5 km. Alphanumeric labels refer to: [1] perimeter dike and ditch, [2] open water, [3] relict drag-line ditch, [A1A] State Highway A1A, and [4] boundaries of enlargement shown in right panel. *Right:* Enlargement of study area indicating approximate location and size of control (squares) and impact (circles) plots. Each plot is 5 × 10 m and consists of a mature, monospecific red mangrove (*Rhizophora mangle* L.) fringe forest that has developed on the perimeter dike's intertidal substrate.

Table 2. Estimate of percent canopy removed from four impact plots. Data generated by post-acquisition processing of Leaf Area Index 2000 Analyzer output. Initial canopy alteration shown in center column; final reduction in right column. Pruning guidelines (described in text) require ≤50% of the canopy remain intact. See Figure 2 for plot locations

Plot	April 19 & 20	May 19
B	34%	45%
D	36%	48%
F	41%	41%
H	21%	47%

terway could be observed. This type of mechanical alteration deviated significantly from standard methods because it required pruning on 1. trees with a DBH >30 cm and 2. branches with a diameter >2.5 cm.

DEP guidelines for selective trimming also mandated at least 50% of the impacted canopy remain intact. In this experiment, a Leaf Area Index 2000 Analyzer (LAI) was used to estimate canopy closure (see Welles, 1990). LAI measurements are based upon the difference between diffuse radiation above (or outside) and below (or inside) the canopy. The above canopy measurements (n=1) were made at an unobstructed site on the perimeter dike (Figure 2) in close proximity to each experimental plot. Below canopy measurements (n=10) were obtained ~30 cm above each litter basket. Prior to selective pruning, LAI measurements of unaltered canopy were obtained from control and impact plots (cf. Table 2). Subsequent measurements were then obtained from the impact plots after removal of each lateral branch until canopy gaps visually appeared to have increased by as much as 50% relative to the initial unaltered canopy. Post-acquisition data processing software was used to convert LAI radiation measurements into percent canopy closure estimates.

In our study, laterals of known diameter (but >2.5 cm) were removed by chainsaw and the cuts treated with tar to minimize subsequent tree damage by infestation. LAI estimates computed after the initial pruning indicated between 21% and 41% of the canopy was removed (Table 2). Because permit guidelines allowed as much as 50% of the canopy to be removed, the DEP requested additional pruning be conducted to maximize the potential effects of selective pruning. Therefore, on 19 May 1995 additional laterals were selectively pruned from all impact plots except F (Table 2).

Light intensity

Access to the LAI was limited to the initial two pruning events and therefore subsequent changes to forest canopies were inferred using a LI-COR Photometric Light Meter with two Quantum Sensors (LI-COR). Like the LAI, LI-COR measurements are used to estimate canopy closure, although the optical sensors and post-acquisition data processing software are not as sophisticated. Prior to selective pruning, simultaneous light measurements were obtained from the control and impact plots using both LAI and LI-COR methods. Simultaneous measurements were also obtained before and after pruning each lateral branch within the four impact plots. When using the LI-COR, data were obtained by placing one sensor at a fixed position outside of the canopy and another ~30 cm above each litter basket. These data are expressed as the percentage of ambient light transmitted through the canopy.

The relationship between the LAI and LI-COR estimates of canopy closure was determined using linear regression analysis of all data collected simultaneously in association with the two selective pruning events. This relationship is expressed by Equation (1):

$$y = -0.07x + 4.08, \qquad (1)$$

where the LAI estimate is represented by the variable y and x is the LI-COR estimate. With this statistically significant relationship established (see Figure 3), percent canopy closure could be estimated from LI-COR measurements of subcanopy light transmission. LI-COR measurements were also recorded once per month throughout the duration of the study (see also Teas, 1979; Snedaker et al., 1992).

Litter fall

Litter fall methods followed the recommendations of Pool et al. (1975), Proctor (1983), Brown (1984) and Twilley et al. (1986). Material was collected at two-week intervals for 21 months and then, after selective pruning, collection continued for an additional 12 months (August 1993 to June 1996). This ~3 year collection interval appears sufficient to assess the effects of selective pruning on litter fall because it is sufficient to capture 1. at least one *Rhizophora mangle* L. reproductive cycle (~16 months; Gill & Tomlinson, 1971) and 2. fluctuations between growth predominance of vegetative and reproductive biomass (~2–3 years; Clough & Attiwill, 1982).

To collect litter fall, wooden baskets (0.25 m^2) were constructed using pressure treated lumber and 1.0 mm^2 fiberglass mesh. Following the recommendations of Newbould (1967) and Brown (1984), ten baskets were placed approximately 1 m above the forest floor at random locations in each plot following the quadrant method of Lahmann (1988).

Captured material was returned to the laboratory, dried to a constant weight at 105 °C and separated into one of four litter fall components 1. leaf, 2. fruit, 3. wood, and 4. detritus (Proctor, 1983; Brown, 1984). The data are expressed in grams of organic matter m^{-2} d^{-1}.

Impact assessment

Canopy alteration and recovery

Successful collection of LI-COR light transmission data was frequently (~30%) inhibited by equipment failure or logistical problems with the staffing of field personnel. These data (Figure 3) were subjected to a power test (α = 0.05) which indicated a sufficient number of measurements were obtained to quantify subcanopy light transmission. The results reveal a relatively constant light level in both control and unaltered impact plots. This clearly reflects the 1. 'ever-growing' characteristics of mangrove described by Gill & Tomlinson (1971) and 2. observations of Whittaker (1966), who suggested canopy closure and light transmission are not strongly correlated with the highly seasonal production of litter fall.

Gaps in the canopy (Welles, 1990) were significantly enhanced by selective pruning and this resulted in an increase in subcanopy light transmission (more than 30%; Figure 3). This increase is analogous to the natural impact of catastrophic events (i.e. hurricanes, freezes) or the localized effects of tree fall and lightning strikes (cf. Lugo & Snedaker, 1974; Putz & Chan, 1986; Snedaker et al., 1992). One obvious positive effect of a mechanically induced increase in subcanopy light transmission is the potential for enhanced seedling recruitment. This response has

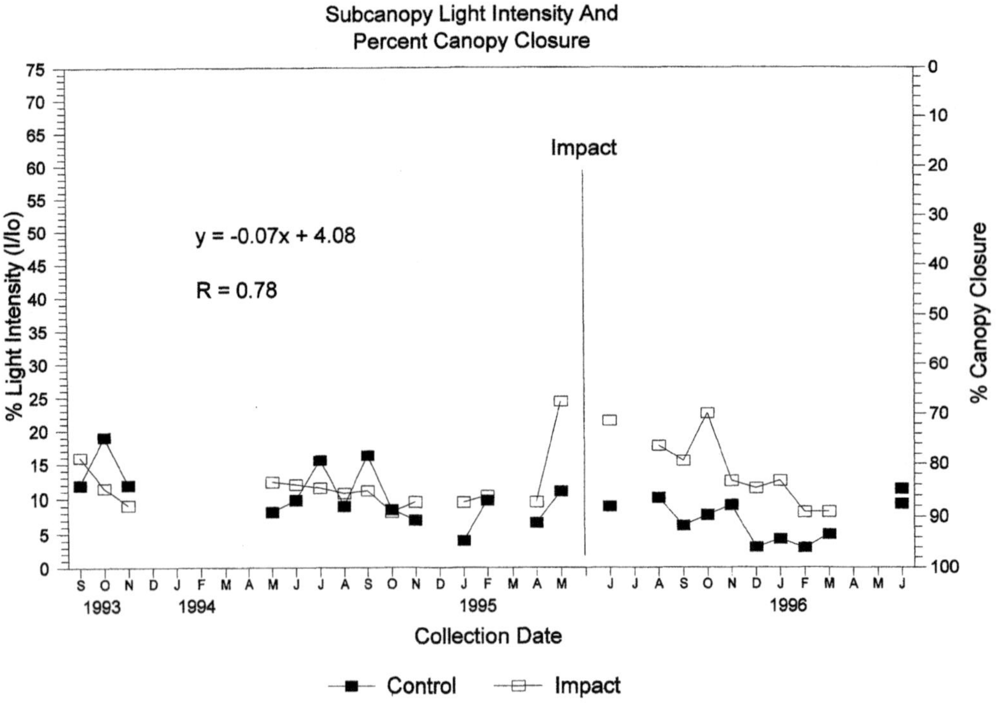

Figure 3. Light intensity data. Although subcanopy light transmission could not be measured ~30% of the time, existing data are sufficient to suggest seasonal variation is not statistically significant. Hence, leaf standing stock remained relatively constant throughout the year.

been observed within the four impact plots and reported by Lahmann (1988). However, several exotic species, including Brazilian Pepper, are now successfully expanding beneath canopy gaps. Subcanopy light transmission within the impact plots has steadily declined since pruning (Figure 3) and by June 1996 mean values had approached those measured in the control plots. The long-term (3–5 yrs) recruitment success of seedlings and invasive exotic species is therefore uncertain.

Litter fall production

The 'ever-growing' and 'ever-flowering' mangrove characteristics reported by Gill & Tomlinson (1969, 1971) is clearly evident in the litter, leaf and fruit fall data collected during this study (Figures 4–6). But a seasonal pattern is also very apparent. Vigorous shoot growth and development from apical vegetative buds during the summer months yields an abundance of leaf material, which rapidly matures and falls toward the forest floor. Since the rates of leaf production and fall both reach maximum levels during the summer months (Gill & Tomlinson, 1971), canopy standing stock remains unchanged (Gill & Tomlinson, 1971; Clough & Attiwill, 1982). Levels of subcanopy light transmission confirm this dynamic equilibrium (Figure 3). Peak fruit (flowers, fruit, propagules) fall is coincident with leaf fall (October and November; Figures 5 & 6) and generated by an increase in both seasonal flower production and the loss of mature propagules. This interval of peak fruit production occurred approximately two months after that reported from the Miami area and nicely illustrates the inverse relationship between temperature and duration of the mangrove reproductive cycle (Gill & Tomlinson, 1971).

Wood, the third most abundant component of litter fall (Table 3), is the only material shed for the forest that does not follow a well defined seasonal pattern (Figure 7). In this study, the release of wood from the canopy is related to wind intensity, which may disrupt the forest during either cyclonic events or winter northeasters (see also Twilley et al., 1986).

About 6% of the litter fall could not be identified and was placed in the detritus category (Table 3). Inspection of the temporal pattern in detrital production (Figure 8) reveals seasonal and event input, clearly reflecting the source of this material is either leaf and fruit or wood, respectively.

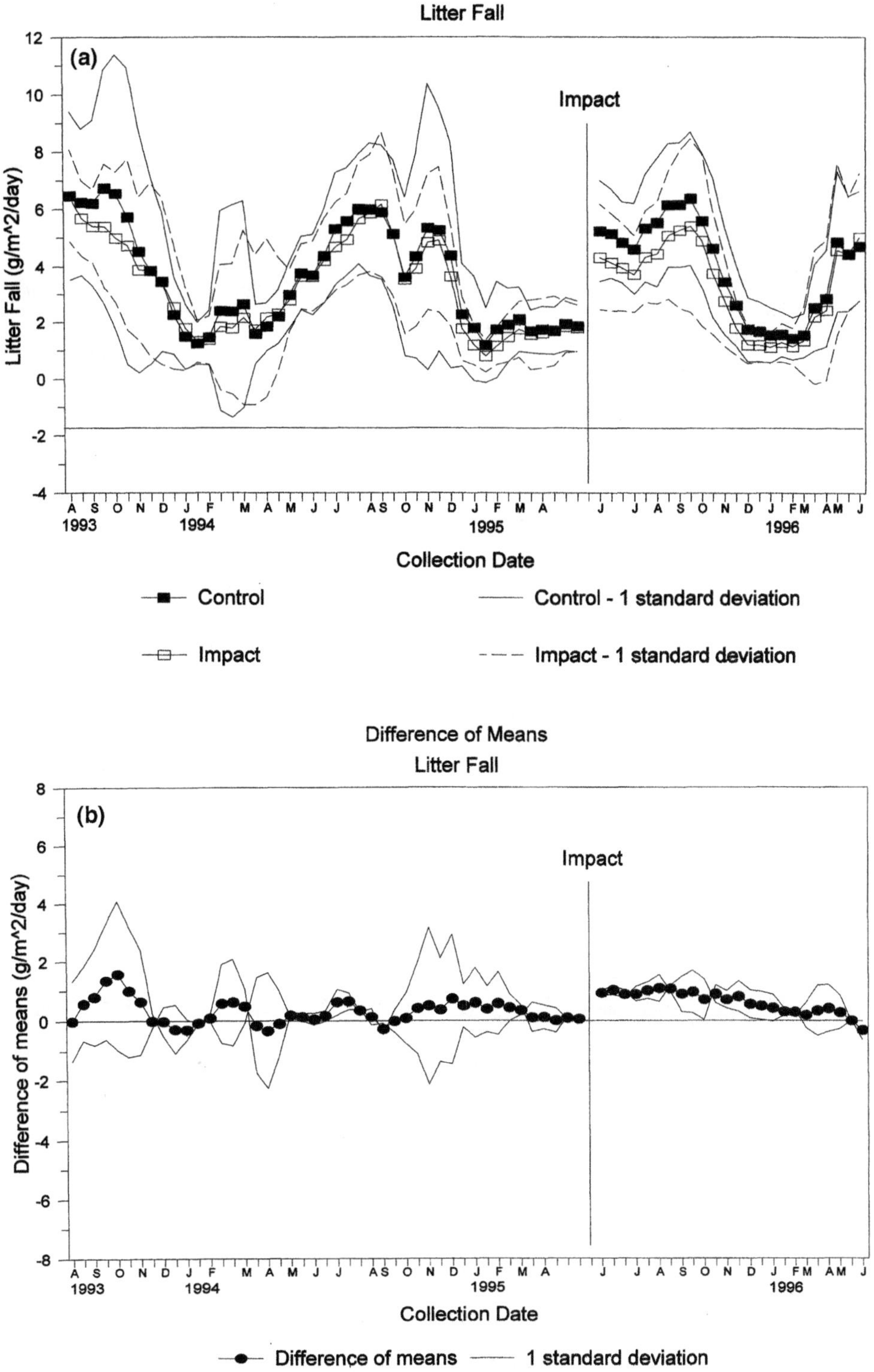

Figure 4. Litter fall data. (A) highly seasonal litter fall, with elevated production between May and December and minimal production between January and April. High production in February 1994 reflects the contribution of a discrete wood fall event generated by strong northeastern winds (see Figure 7). (B) difference between mean litter fall data obtained from control and impact plots before and after trimming.

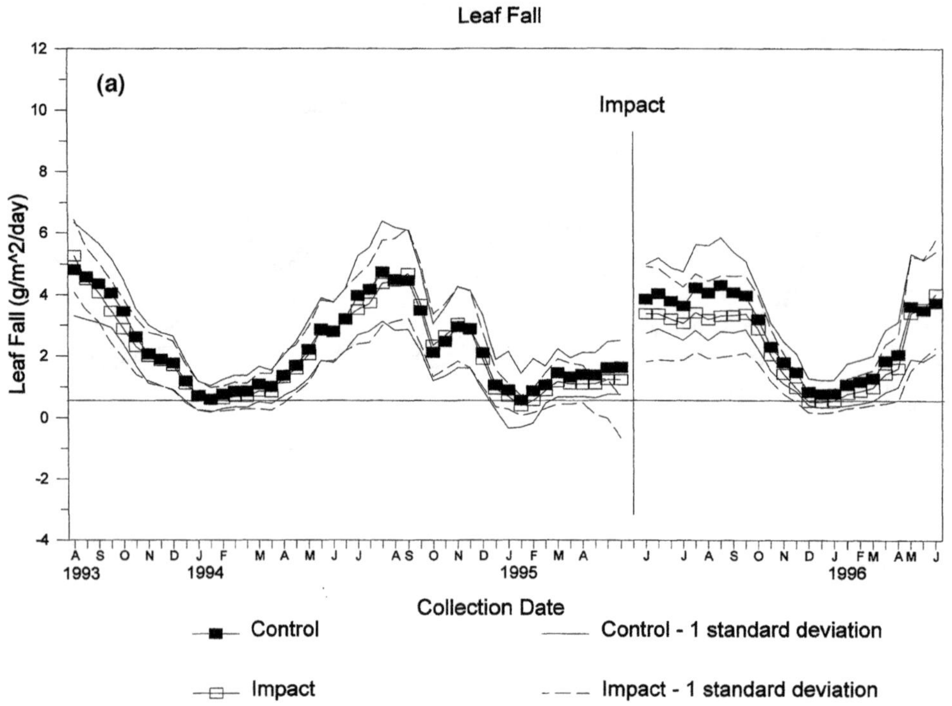

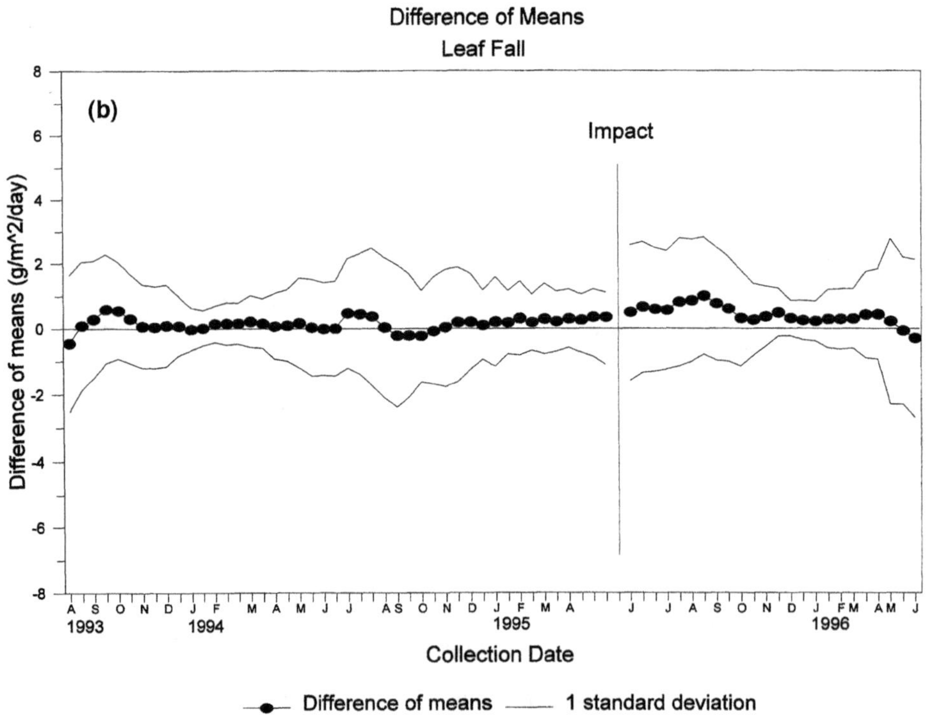

Figure 5. Leaf fall data. (A) a majority of the total litter fall was leaf material (62%) and therefore, seasonal productivity patterns are also apparent in this data set. (B) difference between mean leaf fall data obtained from control and impact plots before and after trimming.

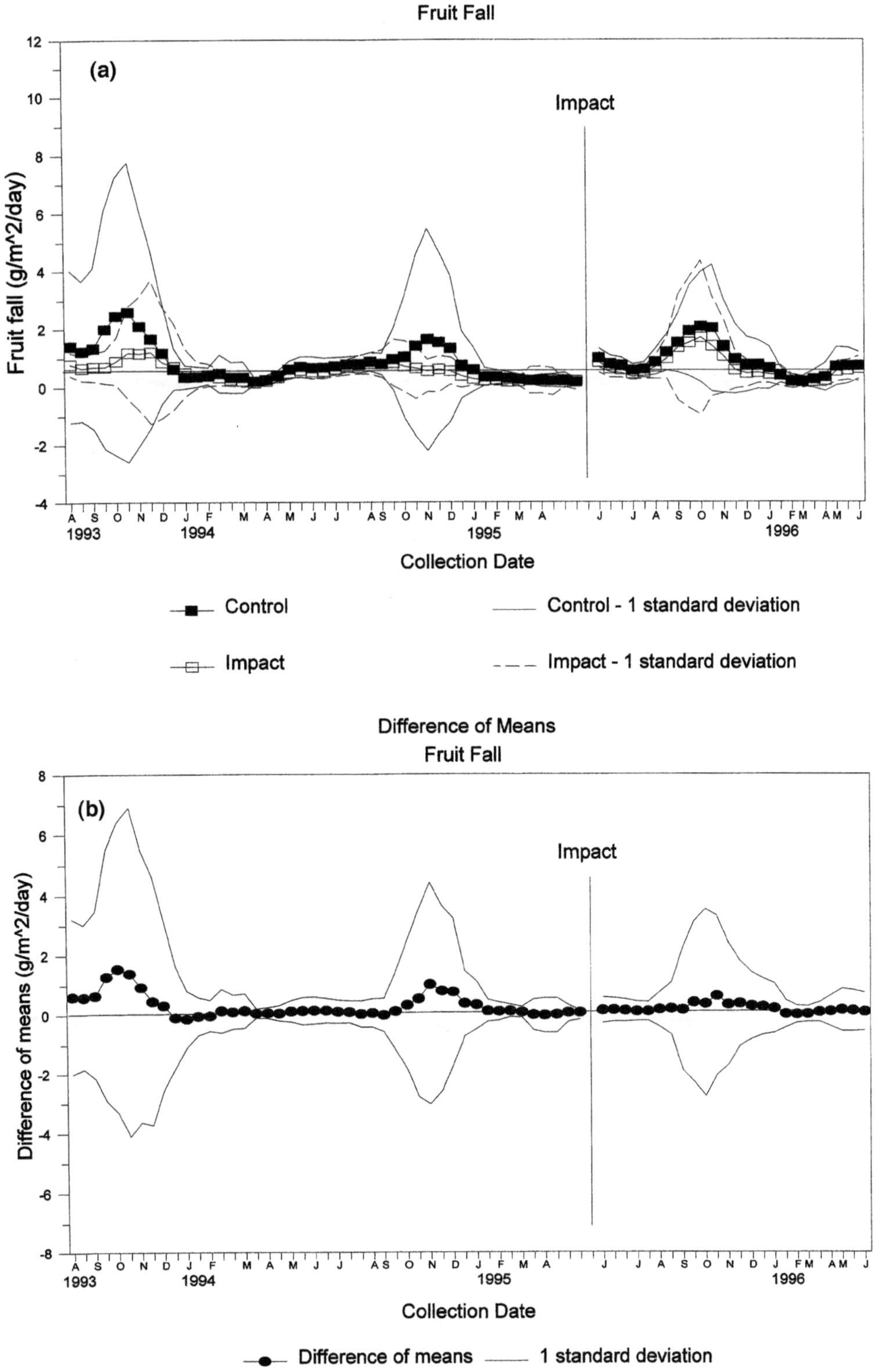

Figure 6. Fruit fall data. (A): fruit fall was the second most abundant component of litter fall (24%) and highly seasonal, although peak production occurred about two months after maximum leaf fall (October through December). (B) the difference between control and impact plot production of fruit before and after trimming.

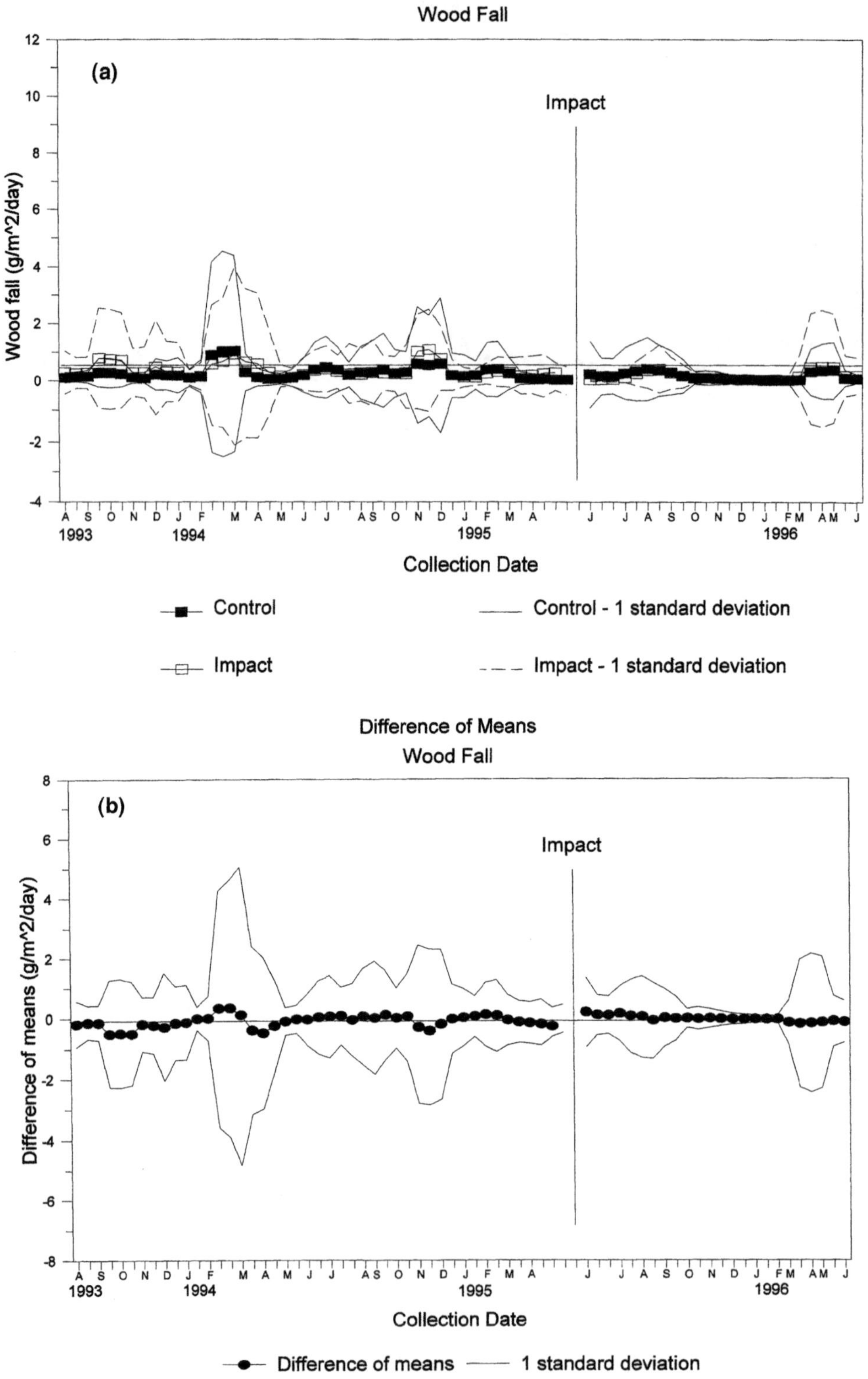

Figure 7. Wood fall data. (A) wood production was the third most abundant constituent of litter fall (10%). Intervals of high production were observed during the pre-trim monitoring and in association with strong winds. (B) The difference between wood fall measured in control and impact plots before and after trimming.

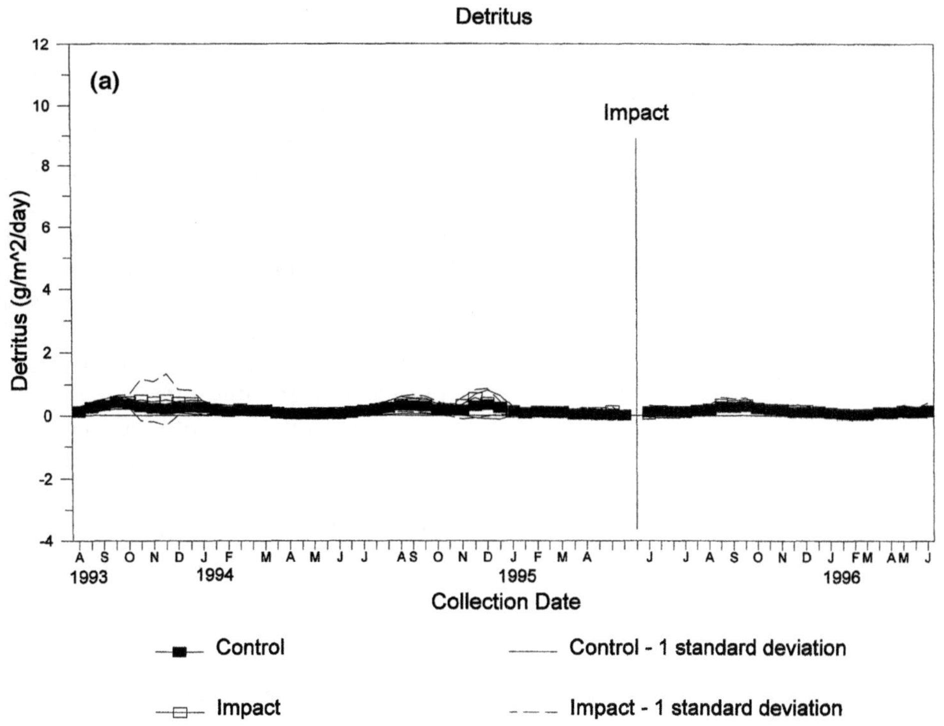

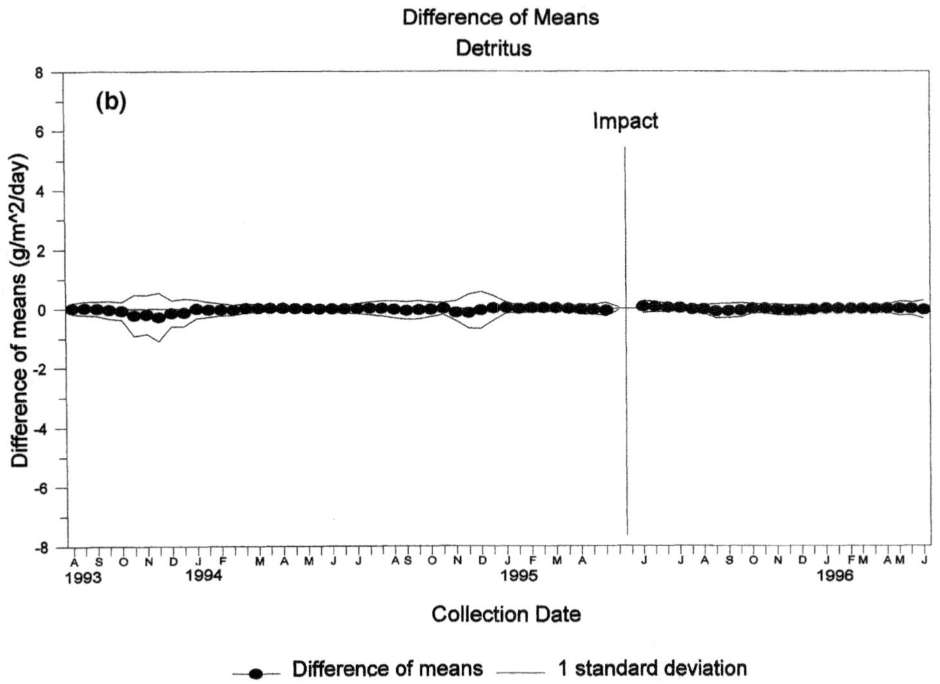

Figure 8. Detritus data. (A) about 6% of the litter fall could not be identified. This material was placed in the detritus category. (B) the difference between control and impact plot production of detritus before and after trimming.

Table 3. Descriptive statistics (α=0.05) calculated using litter fall data collected 21 months before and 12 months after pruning. The percentage of total production is shown adjacent to each litter fall component

Category	Mean		Difference
	Control	Impact	(Control – Impact)
Litter fall			
Before	3.73 ± 2.31	3.23 ± 1.81	0.50
After	3.73 ± 1.96	3.05 ± 1.67	0.68
Leaf fall (61%)			
Before	2.29 ± 1.50	.05 ± 1.39	0.24
After	2.49 ± 1.47	2.04 ± 1.26	0.45
Fruit fall (24%)			
Before	0.89 ± 0.74	0.53 ± 0.32	0.36
After	0.90 ± 0.65	0.62 ± 0.51	0.28
Wood fall (10%)			
Before	0.36 ± 0.72	0.41 ± 0.54	−0.22
After	0.16 ± 0.15	0.15 ± 0.15	0.01
Detritus (5%)			
Before	0.19 ± 0.13	0.24 ± 0.18	−0.05
After	0.19 ± 0.12	0.24 ± 0.12	−0.05

Table 4. Statistical parameters used in H_O testing. If H_O is accepted, productivity values obtained before and after pruning were not significantly different

Category	Statistical Test	Result	Probability	Power	H_O
Litter fall	*t*-test	−0.0884	0.930	0.05	Accepted
Leaf fall	*t*-test	−2.224	0.030	0.488	Rejected
Fruit fall	Mann-Whitney	795	0.789	N/A	Accepted
Wood fall	Mann -Whitney	892	0.324	N/A	Accepted
Detritus	Mann -Whitney	674	0.064	N/A	Accepted

The results of the BACI impact assessment suggest mean forest litter fall production before and after pruning was not significantly different (Table 4). However, inspection of the litter fall data reveal an obvious decline in impact plot production immediately after pruning (Figure 4). This initial (6 month) reduction is logically related to the removalof mature leaf and propagule standing stock, as well as apical vegetative buds capable of generating leaf fall material within the subsequent six months (Gill & Tomlinson, 1969, 1971). The reduction of canopy standing stock by selective pruning is also revealed by the light transmission data, which increased from 10% to 30% (Figure 3).

By the end of the project monitoring period, 12 months after pruning, the graphical trends in litter fall production reflect convergence between control and impact plots (Figures 4–8). This qualitative analysis is supported by the BACI statistical results (Table 4), which are a time-averaged assessment of pre- and post-impact data sets. Except for leaf fall, there was no significant difference between any of the litter fall components before and after selective pruning. Although leaf fall data are significantly different, the low probability value (Table 4) and graphical trends (Figure 5) suggest this difference is minimal and that a return to similar rates of production is forthcoming. The subcanopy light transmission data (Figure 3) also suggests a convergence between control and impact plots is occurring, presumably by apical bud branching and the growth of new vegetative and reproductive canopy standing stock.

Management implications

This project was designed to assess the impact of a selective pruning event on the litter fall production of a red mangrove forest located within the IRL. This impact was an isolated event applied to the forest immediately preceding the productive summer months. But many forests in peninsular Florida are subjected to multiple pruning events or other forms of repeated mechanical alteration. The cumulative effects of repeated alteration on litter fall production, ecosystem function or habitat value are unknown at present. However, if a single impact can enhance forest stress levels (cf. Cintron & Schaeffer-Novelli, 1982) and trigger a short term (~6 months) decline in productivity (this study), a more obvious decline in forest function and value is expected to accompany a more rigorous pruning program (cf. Beever, 1991).

The probability of mechanical alteration inducing a decline in litter fall production is enhanced in regions where the forest is already subjected to stress created by such conditions as elevated salinity (i.e. basin forest), nutrient enrichment (i.e. proximal to urban or agricultural areas) or flooding (i.e. mosquito control water level management). As a majority of IRL mangrove forests are contained within mosquito control impoundments (Rey et al., 1991), seasonal flooding (summer) and elevated salinity (winter) occur throughout the year. Hence, forest litter fall production, habitat value and ecosystem function may already be compromised by management-induced stress enhancement. In other areas (i.e. fringe-forests), greenhouse-induced accelerated sea-level rise may be exerting stress on mangrove forests which have otherwise persisted under relatively stable sea-level conditions for the past 3000 yrs (Parkinson et al., 1994). In summary, the mangrove forests of the IRL may be especially vulnerable to the effects of pruning, particularly if they are subjected to repeated alteration.

This study has also documented a lengthened reproductive cycle in the temperate forests of central Florida (see also Twilley et al., 1986) relative to the subtropical forests of south Florida. Peak fruit production occurs approximately two months after that reported from the Miami area (Gill & Tomlinson, 1971). This mangrove phenology must be taken into consideration when attempting to develop a statewide protection policy. Such a policy must be designed to minimize the impact of mechanical alteration on forest function and value in all climatic zones, while still providing the homeowner with a reasonably simple set of pruning guidelines.

Finally, most impact assessments of mechanical alteration have been conducted using litter fall; one component of above-ground primary production. Yet some studies have suggested below-ground primary production can exceed above-ground production by as much as 50% (cf. Christensen, 1978). While this type of primary production is not a significant component of the detrital-based estuarine food web, it does influence sediment accretion rates, which in turn can alter the frequency of forest flooding (i.e. hydroperiod). If pruning reduces below-ground production of biomass, rates of vertical accretion may decrease, leading to increased forest flooding, enhanced stress and a continuing cycle of declining primary production (both above- and below-ground). In light of the fact that predictions of global climate change include accelerated rates of sea-level rise, the potential impact of declining below-ground primary production could become significant with respect to the long-term sustainability of mechanically altered mangrove forests. Estimates of below-ground biomass production have remained enigmatic for some time, none the less they appear to be yet another important aspect of impact assessment that must be quantified if the long-term effects of mechanical alteration are to be reliably estimated.

Acknowledgements

A majority of the data used in this investigation were obtained by MP-B while pursuing a graduate degree at Florida Institute of Technology, Melbourne, Florida. The following organizations provided funds and support necessary to complete this project: Florida Department of Environmental Protection (in particular, Mr Brian Poole, Mr Rich Mermer and Mr Jeff Beal), IRL National Estuary Program (in particular Mr Bob Day), St. Johns River Water Management District, South Florida Water Management District, St. Lucie County Mosquito Control District (especially Mr James David), Florida Institute of Technology, and the United States Fish and Wildlife Service (in particular Mr Kalani Cairns). Walter Nelson and Hillary Swain are gratefully acknowledged for their guidance in the use of the statistical methods, although we are solely responsible for the interpretations drawn from the output. An initial draft of this manuscript benefitted from the reviews of Kalani Cairns, Brian Poole and Hilary Swain. Anita Bromberg, John Eckert, Lonnie Mowell, George Wiegman and especially Elisa Laurent are also gratefully acknowledged for technical support. Finally, the authors benefitted enormously from the discussions held during meetings of the Florida Department of Environmental Protection's Mangrove Technical Advisory Committee, 1993–1995. In particular Mr James Beever, Mr Robin Lewis, and Dr Eric Heald. RWP dedicates this effort in memory of Dr Heald, whose friendship and myriad contributions toward developing a sustainable future for Florida's vanishing mangrove ecosystems was outstanding.

References

Beever, J. W., 1991. The effects of fringe mangrove trimming for view in the Southwest Florida Aquatic Preserves. Unpublished report. Florida Freshwater Fish and Game Commission, Punta Gorda, Florida: 72 pp.

Brown, M. S., 1984. Mangrove litter production and dynamics. In Snedaker, S. C. & J. G. Snedaker (eds), The Mangrove Ecosystem: Research Methods, Monographs on Oceanographic Methodology, Volume 8. Unesco, Paris: 231–238.

Bunt, J. S., 1982. Studies of mangrove litter fall in tropical Australia. In Clough, B. F. (ed.), Structure, Function and Management of Mangrove Ecosystems in Australia. Australia National University Press, Canberra: 223–237.

Christensen, 1978. Biomass and primary production of *Rhizophora apiculata* Bl. in a mangrove in southern Thailand. Aquat. Bot. 4: 43–52.

Cintron, G. & Y. Schaeffer-Novelli, 1982. Mangrove forests: ecology and response to natural and man-induced stresses. In Coral Reefs, Seagrass Beds, Mangroves and Their Interaction in the

Coastal Zones of the Caribbean. Workshop West Indies Laboratory. Fairleigh Dickinson University, St. Croix, USVI.

Cintron, G. & Y. Schaeffer-Novelli, 1984. Methods for studying mangrove structure. In Snedaker, S. C. (ed.), The Mangrove Ecosystem: Research Methods, Monographs on Oceanographic Methodology, Volume 8. Unesco, Paris: 91–113.

Clough, B. F. & P. M. Attiwill, 1982. Primary productivity in mangroves. In Clough, B. F. (ed.), Structure, Function and Management of Mangrove Ecosystems in Australia. Australia National University Press, Canberra: 213–222.

Day, J. W., R. H. Day, M. T. Barreiro, F. Ley-Lou & C. J. Madden, 1982. Primary production in the Laguna de Terminos, a tropical estuary, in the southern Gulf of Mexico. Oceanol. Acta SP: 269–276.

Day, J. W. Jr., W. H. Connor, R. H. Ley-Lou, R. H. Day & A. M. Navarro, 1987. The productivity and composition of mangrove forests, Laguna de Terminos, Mexico. Aquat. Bot. 27: 267–284.

Estevez, E. D. & R. K. Evans, 1978. The effects of pruning on production by the red mangrove *Rhizophora mangle*. In Leslie, R. R. (ed.), Proceedings Fifth Annual Conference; Restoration of Coastal Vegetation in Florida. Hillsborough Community College. Tampa, Florida: 26–42.

Gill, A. M. & P. B. Tomlinson, 1969. Studies on the growth of red mangrove (*Rhizophora mangle* L.). Habitat and general morphology. Biotropica 1: 1–9.

Gill, A. M. & P. B. Tomlinson, 1971. Studies on the growth of red mangrove (*Rhizophora mangle*). Phenology of the shoot. Biotropica 3: 109–124.

Jordan, C. F., 1971. A world pattern in plant energetics. Am. Scient. 59: 425–433.

Lahmann, E. J., 1988. The effects of different hydrological regimes on the productivity of *Rhizophora mangle*. A case study of mosquito control impoundments of Hutchinson Island, St. Lucie County, Florida. Unpublished dissertation. University of Miami, Coral Gables, Florida: 149 pp.

Lewis, III, R. R., R. G. Gilmore, D. W. Rewz & W. E. Odum, 1985. Mangrove habitat and fishery resources in Florida. In Seaman, W. (ed.), Florida Aquatic Habitat and Fishery Resources. Florida Chapter of American Fisheries Society: 28–336.

Lugo, A. E. & S. C. Snedaker, 1974. Properties of mangrove forests in southern Florida. In Walsh G., S. Snedaker & H. Teas (eds), Proceedings of the International Symposium on Biology and Management of Mangroves, Vols. I and II. Institute of Food and Agriculture Science, University of Florida, Gainesville (FL U.S.A.): 170–212.

Newbould, P. J., 1967. Methods for estimating primary production of forests. IBP Handbook No. 2. Blackwell Scientific, Oxford: 62 pp.

Odum, W. E. & E. Heald, 1972. Trophic analysis of an estuarine mangrove community. Bull. Mar. Sci. 22: 671–732.

Odum, W. E., C. C. McIvor & T. J. Smith III, 1985. The ecology of the mangroves of south Florida: a community profile. U.S. Fish and Wildlife Service, Office of Biological Services, Washington (D.C.): 144 pp.

Olson, J. S., 1963. Energy storage and the balance of producers and decomposers in ecological systems. Ecology 44: 322–331.

Parkinson, R. W., R. D. DeLaune & J. R. White, 1994. Holocene sea-level rise and the fate of mangrove forests within the wider Caribbean region. J. Coast. Res. 10: 1077–1086.

Pool, D. J., A. E. Lugo & S. C. Snedaker, 1975. Litter production in mangrove forests of southern Florida and Puerto Rico. In Walsh, G., S. Snedaker & H. Teas (eds), Proceedings of the International Symposium on Biology and Management of Mangroves, Vol. I and II. Institute of Food and Agricultural Science, University of Florida, Gainesville (FL U.S.A.): 213–237.

Proctor, J., 1983. Tropical litter fall I: problems of data comparison. In Sutton, S. L., T. C. Whitmore & A. C. Chadwick (eds), Tropical Rainforests: Ecology and Management. Blackwell Scientific Publishing, Oxford: 267–273.

Putz, F. E. & H. T. Chan, 1986. Tree growth, dynamics, and productivity in a mature mangrove forest in Malaysia. Forest Ecol. Mgmt. 17: 211–230.

Rey, J. R., T. Kain & R. Stahl, 1991. Wetland impoundments of east-central Florida. Florida Scientist 54: 33–40.

Snedaker, C. S., M. S. Brown, E. J. Lahmann & R. J. Araujo, 1992. Recovery of mixed-species mangrove forests in south Florida following canopy removal. J. Coast. Res. 8: 919–925.

Stewart-Oaten, A., W. W. Murdoch & K. R. Parker, 1986. Environmental impact assessment: 'pseudoreplication' in time?. Ecology 67: 929–940.

Teas, H. J., 1979. Silviculture with saline water. In Hollaender, A., J. C. Aller, E. Epstein, A. S. Pietro & O. R. Zabrosky (eds), The Biosaline Concept. Plenum Publishing Corporation, New York: 117–161.

Tomlinson, P. B., 1986. Botany of mangroves. Cambridge University Press, Cambridge: 413 pp.

Twilley, R. R., A. E. Lugo & Patterson-Zucca, 1986. Litter production and turnover rate in basin mangrove forests in southwest Florida. Ecology 67: 670–683.

Welles, J. M., 1990. Some indirect methods of estimating canopy structure. Remote Sensing Reviews 5: 31–43.

Whittaker, 1966. Forest dimensions and production in the Great Smokey Mountains. Ecology 47: 103–121.

Hydrobiologia **413:** 77–88, 1999.
R.S. Dodd (ed.), Diversity and Function in Mangrove Ecosystems.

Breakdown of mangrove leaf litter in a managed mangrove forest in Peninsular Malaysia

E. C. Ashton, P. J. Hogarth & R. Ormond
Tropical Marine Research Unit, Department of Biology, University of York, P.O. Box 373, York YO10 5YW, U.K.

Key words: decomposition, litter bags, diversity, ecosystem function

Abstract

Decomposition of *Rhizophora apiculata*, *Rhizophora mucronata*, *Bruguiera parviflora* and *Sonneratia alba* leaves was studied *in situ* using litter bags in both Cleared and Virgin Jungle Reserve (VJR) mangrove forests in Peninsular Malaysia. A single exponential model best described the rate of decomposition for each species. All leaf species decomposed faster in the VJR site than in the Cleared site (*R. apiculata* $P < 0.05$, *R. mucronata* $P < 0.01$, *B. parviflora* $P < 0.01$, *S. alba* not significant and mixed bags $P < 0.05$). The rate of decomposition was species specific: *Sonneratia alba* leaves decomposed the fastest ($P < 0.001$) in both sites. The time in days required for the loss of half the initial dry mass (t_{50}) was Cleared site: *R. apiculata* 76, *R. mucronata* 122, *B. parviflora* 122, *S. alba* 22, mixed 51; VJR: *R. apiculata* 43, *R. mucronata* 34, *B. parviflora* 70, *S. alba* 15 and mixed 32. Increasing litter diversity, by mixing leaves of different species in bags, had no effect on decomposition rate. The mass of air controls showed an initial decline to 65% in 14 d but then remained fairly constant (t_{50}=108 d). This initial loss may represent the leaching of dissolved organic matter. Water controls (mixed litter bags submerged in seawater) had a t_{50} of 10 d, a rate significantly different ($P < 0.01$) from air controls. Our results show that breakdown of leaf litter is site and species dependent. This affects ecological functioning of the mangrove ecosystem and may have implications for management and conservation of mangroves.

Introduction

Mangrove forests are productive ecosystems and support a high abundance and diverse variety of wildlife (Ong, 1995). Aksornkoae (1986) suggests this is as a result of high leaf production, leaf fall and rapid breakdown of the detritus. Litterfall has been estimated to account for 30–60% of total primary production (Bunt et al., 1979). The importance of mangrove leaf litter in the maintenance of detrital-based food webs in the coastal environment and their significance for coastal fisheries has been indicated for some time (Golley et al., 1962; Odum & Heald, 1975; Ong et al., 1984; Lee, 1995). However, mangrove forests are disappearing at alarming rates due to various human impacts (reclamation for aquaculture, farming, residential and industrial development), the greatest losses occurring in south-east Asia. Between 1980 and 1990, Malaysia lost 12% of its mangrove forests, equivalent to an annual loss of 1.3% (Spalding et al., 1997). In order to conserve and sustainably manage the mangrove ecosystem, it is necessary to understand the key processes of production and breakdown of mangrove litter.

Leaf breakdown is defined as weight loss due to physical fragmentation (caused by abiotic factors), animal feeding, microbial activity and leaching (Stewart & Davies, 1989). The chemistry of mangrove detritus changes profoundly during decomposition and aging (Rice & Tenore, 1981). When leaves fall into streams and are wetted, there is initially a very rapid leaching of dissolved organic matter (DOM), including organic carbon, nitrogen (Fell et al., 1975; Newell et al., 1984; Robertson, 1988) and tannins (Steinke et al., 1993a,b). The percentage mass loss of the leaf material leached varies according to the leaf species and site (Robertson et al., 1992). Recorded losses range from 14 to 40% of the initial dry mass of the mangrove leaves (Van der Valk & Attiwill, 1984; Camilleri & Ribi, 1986; Steinke et al., 1993a). Leaching occurs over a period of 3–28 d (Cundell et al., 1979; Steinke et al., 1993a).

The remaining leaf biomass, particulate organic matter (POM), is decomposed more slowly and is highly dependent on the action of bacterial and fungal communities, which develop rapidly on leaves (Schleyer, 1986; Neilson & Richards, 1989; Steinke et al., 1990; Robertson et al., 1992) and which are thought to increase the absolute nitrogen content of the leaf litter (Fell et al., 1975; Rice & Tenore, 1981). The microflora condition the leaf litter for invertebrates to then feed on it. Fragmentation due to non-biotic forces may also be significant in leaf breakdown.

Diverse communities possess greater structural and functional versatility than individual species, and this may enable them to exploit environmental resources more completely (Jolliffe, 1997). Also, greater species diversity may help to sustain productivity by stabilizing community performance under varying environmental conditions (McNaughton, 1993). Evidence from recent studies suggests that productivity can be enhanced in mixed associations (Naeem et al., 1994; Tilman et al., 1996; Jolliffe, 1997; Naeem & Li, 1997). For example, Tilman et al. (1996) present data for a grassland ecosystem and show that mixed plant populations enhance nutrient cycling. However, few studies have looked at the effect of increased litter diversity, the principal examples are Mall et al. (1991), Wardle et al. (1997) and Hooper & Vitousek (1998). Since leaf breakdown plays a key role in ecosystem function, species richness of leaf litter may be important in determining the nature of relationships between biodiversity and ecosystem properties (Wardle et al., 1997). It is essential to understand the significance of the biodiversity of mangrove ecosystems for their function and stability. The rates of leaf breakdown were, therefore, determined for four mangrove species. The effect of increasing the diversity of leaf litter was assessed by combining all four species together in mixed litter bags and comparing the rate of breakdown with that predicted from knowledge of the rates for each of the four constituent species separately.

Management of mangroves has generally resulted in monoculture stands of *Rhizophora* spp. (Gong & Ong, 1995). A good example of a managed mangrove forest is the Matang Mangrove Forest Reserve, Peninsular Malaysia, which has been intensively managed for the production of fuelwood, charcoal and poles on an apparently sustainable basis for 90 years. *Rhizophora* spp. are planted for their productivity and quality of charcoal wood. The trees are thinned after 15 and 20 years for poles and are felled after 30 years (Gan, 1995). However, there is concern that timber production has slowly declined under this regime (Haron, 1981; Gong et al., 1984; Gan, 1995). This situation provides an opportunity and a rationale for investigating whether community ecological function, including productivity and detritus processing are reduced in mangrove stands under different management.

The null hypotheses tested in this study were: 1. there is no difference in rate of decomposition between a Cleared mangrove site and an unmanaged site, a Virgin Jungle Reserve (VJR); 2. there is no difference in rate of decomposition between mangrove leaf species; and 3. higher diversity of leaf species has no effect on the rate of breakdown of the leaves.

Materials and methods

Study sites

This study was undertaken in the Matang Mangrove Forest Reserve; a continuous belt of mangrove forest, with an area of 40 151 ha, located (latitude 4° 15′ N–5° 1′ N and longitude 100° 2′–100° 45′E) on the north-west coast of Peninsular Malaysia (Figure 1). Experiments were carried out both in a recently clear-felled, and yet to be replanted plot (Cleared), and in the unmanaged Pulau Kecil Virgin Jungle Forest Reserve (VJR), which is a mixed species area that has been protected for 80 years (Figure 1).

The clear-felled plot, cleared less than one year previously, comprised a few *Rhizophora apiculata* seedlings and many dead branches and bark. The VJR plot was dominated by mature *Rhizophora apiculata* Blume and *R. mucronata* Lamarck with subdominants of *Bruguiera parviflora* Wight and Arnold ex Griffith and *B. cylindrica* (L.) Bl. There was a thick layer of leaf litter on the forest floor. The two plots were chosen for their proximity to each other as well as the contrasting management regimes. Experiments were carried out within 100 m^2 quadrats placed 20–30 m from the shoreline. *Rhizophora apiculata* and *R. mucronata* are the dominant mangrove species in the Matang study area. *Bruguiera parviflora* and *Sonneratia alba* J. Smith, thought of as pests by foresters, are also abundant. Leaves of these four species were used in this study of the rates of leaf litter processing.

The climate is equatorial with monsoon periods during February–April and September–November. Annual rainfall in the Matang Mangrove Forest Reserve ranges from 2000 to 3000 mm (Gan, 1995). The closest town to the Matang Mangrove Forest Reserve is Taiping, which is 10 km inland and receives

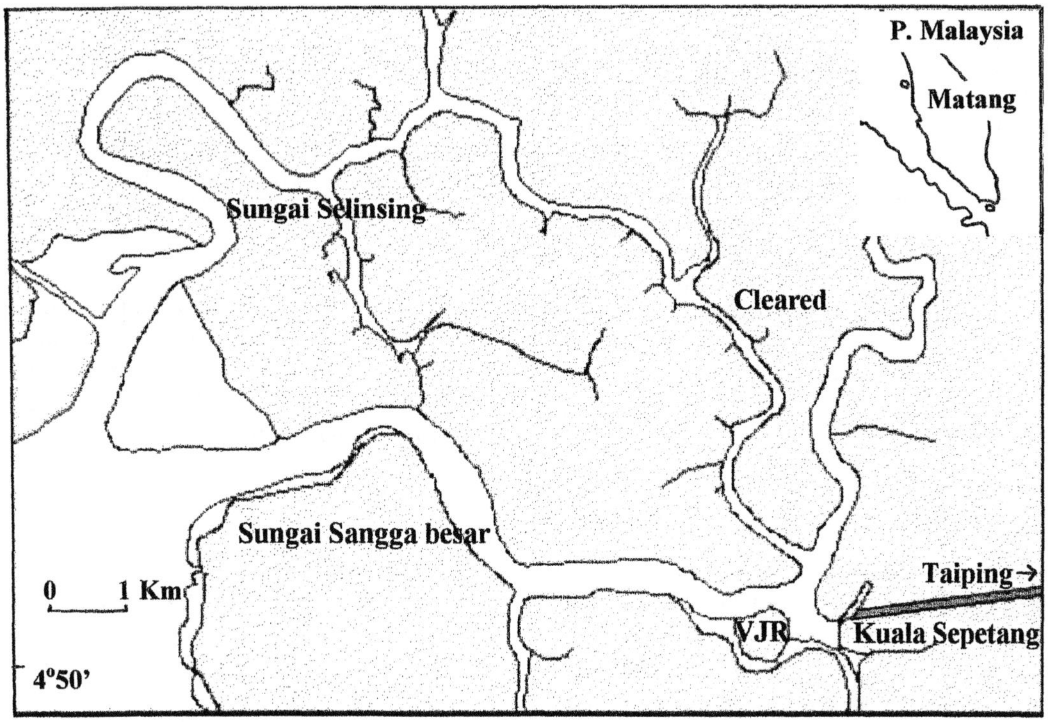

Figure 1. The Matang Mangrove Forest Reserve and the positioning of the Cleared and VJR sites.

the highest rainfall in Malaysia, an annual mean of 4092 mm (mean of 1986–1996). Freshwater flow into the estuary is through numerous streams and ground run-off. The tides are semidiurnal with a pronounced fortnightly neap-spring cycle and a tidal range of 3.3 m.

This study was undertaken between October and December, 1997. In Lubok Merbau (closest principal meteorological station), 262 mm, 216 mm and 80 mm of rain were recorded for the months of October, November and December, with 22, 24 and 16 number of rain days, respectively. The mean monthly humidity was 86.9% in October and decreased to 82.9% in December. Mean maximum temperature was 31.9 °C in October and 32.3 °C in November and December, and mean minimum temperatures 23.0, 23.1 and 22.6 °C for the respective months of the study period.

Environmental conditions (pH, salinity, redox potential, temperature and tidal inundation) were recorded within each quadrat according to the methods described in English et al. (1994). The standard methods of Cintron & Schaeffer-Novelli (1984) were used to determine tree density and basal area, an indicator of tree biomass. Leaf fall was collected during the study period in four 1 m^2 traps placed in a replicate 100 m^2 quadrat positioned 100 m away from the litter bags. The traps were attached to trees and elevated above high tide. Litter fall was collected after 4 and 8 weeks and oven dried to constant weight at 105 °C. The standing crop of leaf litter on the forest floor was collected at the start of the experiment and at 4 and 8 weeks from four permanently marked 1 m^2 quadrats, and oven dried at 105 °C until constant weight. In addition the numbers of leaves on the forest floor were counted in ten replicate 1 m^2 quadrats within each site (standing litter stock).

Leaf decomposition experiments

Litter bags were used to estimate the rates of mangrove leaf breakdown for four species, *Rhizophora apiculata, R. mucronata, Bruguiera parviflora, Sonneratia alba* and mixed bags of all four species. The leaves used were senescent leaves (yellowish in colour) either freshly fallen or easily hand picked from the trees in the surrounding area. Senescent leaves were used instead of fresh because these are the majority of leaves on the forest floor and they have already started their decomposition. These leaves were air dried for 24 h, so that no surface water remained, and then 28.0 g (~8 g of dry mass) were placed into 300 mm × 300 mm nylon mesh bags with a mesh size of 1 mm^2. The mixed bags consisted of 7.0 g of each species. To prevent the leaves from folding and clumping, they were laid flat in the bags.

The mesh bags were securely tied to aerial roots so that they lay flat on the sediment. Five bags of each treatment (four single species and one mixed) were removed from each site at each of 2, 7, 14, 28, 42 and 56 d (i.e. total of 300 bags). These bags were returned to the laboratory where they were gently washed in a sieve to remove sediment. The resulting litter was oven dried to constant mass at 105 °C and the final dry mass recorded.

Control experiments

Two control experiments using mixed litter bags (7.0 g of *R. apiculata, R. mucronata, B. parviflora* and *S. alba*) were also run at the same time to determine the direct effect of exposure to air or water, by leaving litter bags in air for 100% of the time 'air controls', or in water 100% of the time 'water controls'. Air controls of mixed bags were suspended horizontally, 2 m above the ground, at the Virgin Jungle Reserve. Water controls of mixed bags were totally submerged for the whole of the sampling period. Five replicate litter bags were sampled at 14, 28, 42 and 56 d with the exception of the water controls at 56 d, when the remaining bags had disappeared.

Calibration experiments

To estimate the initial dry mass of leaves from fresh mass, senescent leaves of *R. apiculata, R. mucronata, B. parviflora* and *S. alba*, of a range of leaf sizes placed in the bags, were collected and weighed. Leaves were weighed both after air drying for 24 h and after oven drying to constant mass at 105 °C. A plot of dry mass against fresh mass showed significant linear relationships for all the species. A simple linear regression with a line constrained through the origin determined the regression equations for each species, which were then used to estimate the initial dry mass of leaves in the single species and mixed bags.

Statistical analysis

The percentage of initial dry mass remaining in the litter bags was determined and one-way and two-way Analyses of Variance (ANOVA) carried out to investigate the effects of site, species and time. The relationship between percentage dry mass remaining in litter bags and sampling time for all species at both sites best fitted a negative single exponential model:

$$X_t = X_0 \, e^{-Kt},$$

where X_t is the percentage of the initial material X_0 remaining after time t (days) and K is a decay coefficient (d^{-1}). The times required for the decomposition of half the initial material (t_{50}) were determined (t_{50}=ln 2/K). The exponential curves obtained were converted to straight lines ($\log_{10}$) and t-tests between two linear log regressions were carried out for every combination of litter bags. All statistical analyses were performed using SPSS.

Results

The site characteristics relevant to this present study are shown in Table 1.

Calibration experiment

The summary of least squares regressions constrained through the origin, of dry mass on fresh mass for each species are shown in Table 2. The regression slopes were compared by t-test and superscript letters denote results that were not significantly different at the P <0.01 level. *R. apiculata* and *B. parviflora* had significantly different regression slopes (P <0.01) from *R. mucronata* and *S. alba*, but were not significantly different from each other.

Control experiments

The mass of air controls showed an initial sharp decline to 65% of initial mass in 14 d, but then remains virtually constant, giving a relatively poor fit to an exponential model (R^2=0.415) (see Table 3). The rate of mass loss in the water controls is very fast, with a half life (time required for the decomposition of half the initial material), t_{50}, of 10 d (Figure 2; Table 3). A t-test between the two linear log regressions of air and water showed that the difference was highly significant (P <0.01). Both were also significantly different from the experimental bags. Rates of mass loss of water controls were only not significantly different from those for *S. alba* in the VJR. Rates of mass loss of air controls were only not significantly different from those for *B. parviflora* in the VJR and for *R. apiculata, R. mucronata* and *B. parviflora* in the Cleared site.

Table 1. Characteristics of the Cleared and VJR study sites. Mean values are given with the standard errors in parentheses; *n* is the number of replicates

	Unit	*n*	Cleared site	VJR site
Total tree density	No. 100 m^{-2}	1	3	26
Total stand basal area	cm^2 100 m^{-2}	1	283	6216
Leaf litter fall	Dry wt g m^{-2} d^{-1}	8	0	2.79 (0.26)
Standing leaf litter crop	Dry wt g m^{-2} d^{-1}	4, 12	1.51 (0.18)	2.0 (0.2)
Standing leaf litter stock	No. leaves m^{-2}	10	6 (1.3)	43 (7.6)
Redox potential of soil (surface)	mV	6	−73.8 (65.0)	37.5 (44.8)
Redox potential of soil (depth=0.5 m)	mV	6	−255.3 (50.0)	−112.7 (22.1)
Salinity of soil (surface)	ppt	6	17.3 (3.16)	29.5 (0.41)
Salinity of water (depth=0.5 m)	ppt	6	12.17 (1.9)	22.0 (1.32)
Air temperature	°C	6	31.6 (0.53)	29.07 (0.16)
Soil temperature	°C	6	30.53 (0.3)	26.46 (0.04)
Height above chart datum	cm	4	221 (3.46)	283 (2.06)
Flooding frequency	% Flood tides yr^{-1}	4	39	7

Table 2. Regression equations with standard errors of slope in parentheses and regression coefficients (R^2) to determine dry mass (y) from fresh mass (x) for mangrove species. Same letters denote species not significantly different at the $P<0.01$ level

Species	Regression equation	R^2
Rhizophora apiculata[a]	y=0.3184x (0.0032)	0.983
Rhizophora mucronata[b]	y=0.2686x (0.0023)	0.986
Bruguiera parviflora[a]	y=0.3166x (0.0018)	0.997
Sonneratia alba[b]	y=0.2783x (0.0056)	0.976

Leaf decomposition experiments

Analysis of variance showed that there were highly significant differences in percentage dry mass remaining between sites ($P<0.001$) and there was also a highly significant interaction between site and time ($P<0.001$). All the leaf species decomposed faster in the forested mangrove site (VJR) than the Cleared site, although *S. alba* was not significant (*R. apiculata* $P<0.05$, *R. mucronata* $P<0.01$, *B. parviflora* $P<0.01$, *S. alba* P=0.11 and mixed bags $P<0.05$) (Figure 3). However, there was no interaction between site and species. So all the species were affected in the same way. There was no 3-way interaction with site×species×time. The decay constants and times required for the decomposition of half the initial material (t_{50}) are given in Table 3 for species at each site.

The effect of species on percentage dry mass remaining is highly significant (at $P<0.001$) and the interaction of species with time is highly significant (at $P<0.001$). The rate of decomposition was species specific: *Sonneratia alba* leaves decomposed the fastest ($P<0.001$) at both sites with a half life of 15 d at the VJR compared to 43 d for *Rhizophora apiculata* (Figure 3; Table 3). The rate was about the same for *R. apiculata, R. mucronata* and *Bruguiera parviflora.*

The observed results for mixed leaf species bags were compared with the results expected from the combination of all the single species for both sites (Figure 4 & Table 4). There was no significant difference in the rates of leaf breakdown between the observed results for multi species leaf bags and the expected rate at either site when a t-test was performed on the 2 log linear regression lines.

Discussion

From this study, the decay parameters for mangrove litter are shown to be site and species dependent. Leaves in litter bags may face a slightly different microhabitat and microclimate compared to the natural environment, so the results are not the actual litter breakdown in the field, but an indication of potential leaf breakdown (Tam et al., 1990). The litter bag method may underestimate actual decomposition but will reflect trends and allow comparison among species and sites (Wieder & Lang, 1982). The mesh size of the bags allows small invertebrates, fungi and bacteria access to the leaves, but the large invertebrates such as crabs, which are a major factor in leaf de-

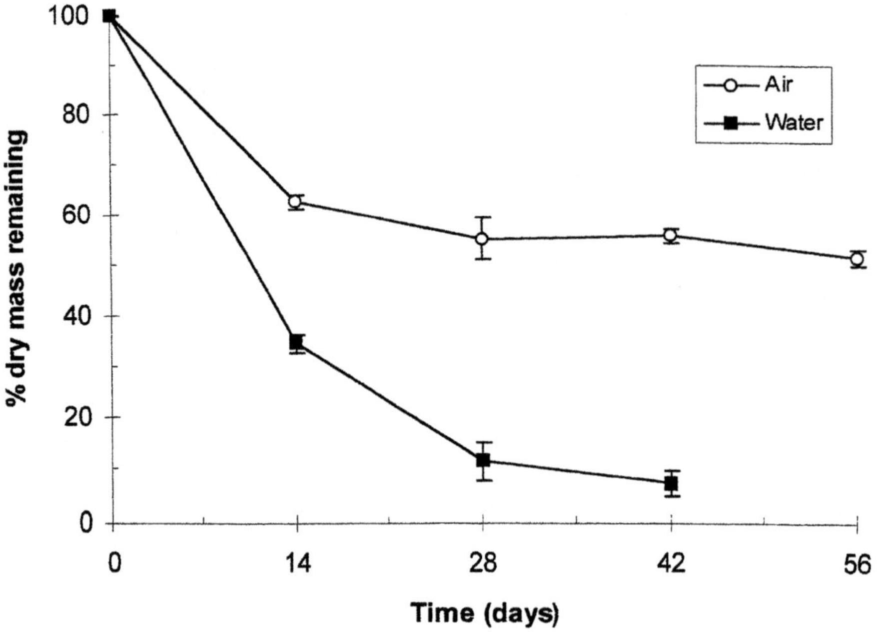

Figure 2. Air and Water Controls: mixed litter bags in air 100% of the time and water 100% of the time. Mean dry mass remaining after various time intervals, as % of initial dry mass.

gradation (Steinke et al., 1993b), are excluded. The role that larger macro-invertebrates play in mangrove leaf breakdown in the Matang was investigated by leaf tethering experiments (Ashton, 1999).

The decay pattern showed rapid early losses in mass followed by a slower decrease for the remainder of the experiment. A double exponential model has been recommended by Wieder & Lang (1982) as most suitable for decomposition data. The *y* intercept for a single exponential curve consistently falls below 100% (see Table 3): this is a consequence of the rapid initial loss of mass due to leaching of dissolved organic matter, and suggests that a double exponential would be a more appropriate description of the data in this study. However, the single exponential model was used for a number of reasons: it is an easily applicable and robust method; decay constants and half-lives can be compared within this study and with other experiments conducted elsewhere; and there are insufficient data points within the first 14 d to determine the leaching exponential curve.

The permanently submerged water controls decomposed most rapidly (t_{50}=10 d compared with 15–122 d for principal treatments). Leaves decompose faster in subtidal regions than in intertidal (Robertson et al., 1992). Water soaking causes leaching of labile materials (Robertson, 1988) and promotes microbial activity (Tam et al., 1990) both of which increase decomposition rates. Also, the tide's physical impact on the leaves may cause fragmentation and thus increased weight loss. Another reason the dry mass loss is so marked may be due to the abundant invertebrate shredders (amphipods) found in the water control litter bags. Mackey & Smail (1996) and Steinke et al. (1983) also comment on the presence of large numbers of isopods and amphipods in litter bags, which must have entered as juveniles through the mesh and were then unable to escape after feeding. Poovachiranon et al. (1986) and Camilleri (1982) suggested that leaves were directly fed on by mangrove amphipods. The density of the macroinvertebrates (Basaguren & Pozo, 1994) influences the rate of decomposition.

The air controls, that were not inundated at all, decomposed the slowest. The leaves are not available to the decomposing leaf fauna and flora and, therefore, their loss in mass is most probably due only to leaching of soluble dissolved organic matter by rainwater. Howard & Howard (1974, cited in Wieder & Lang (1982)) suggested that in the absence of the faunal component, decomposition would not proceed beyond a certain point. This has been confirmed by the air controls up to 56 d.

The Cleared site was inundated by tide more frequently than the VJR; it was covered by 39% of flood

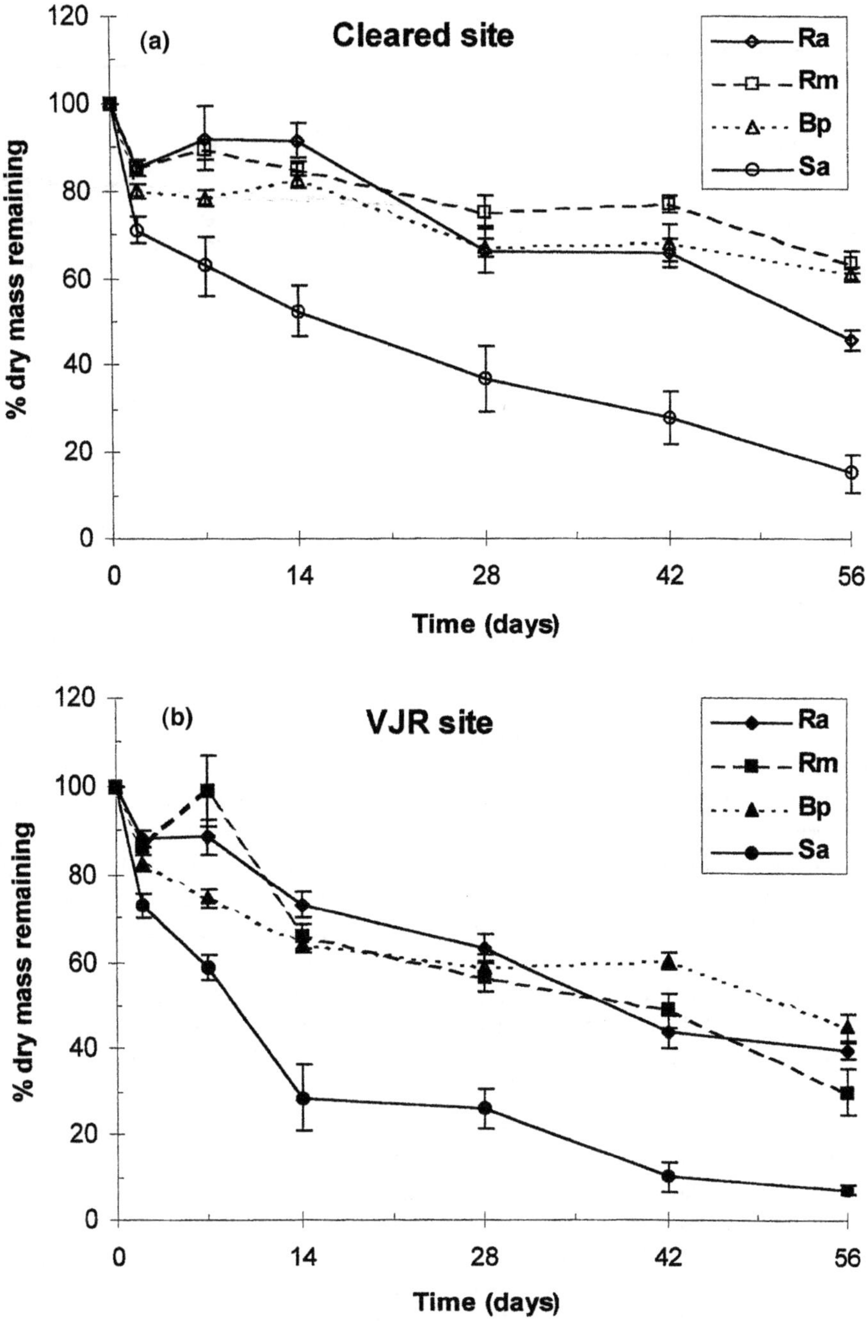

Figure 3. The mean dry mass remaining in litter bags after time, as % of initial dry mass, at (a) the Cleared site and (b) the VJR site for each mangrove species Ra=*Rhizophora apiculata*, Rm=*Rhizophora mucronata*, Bp=*Bruguiera parviflora* and Sa=*Sonneratia alba*.

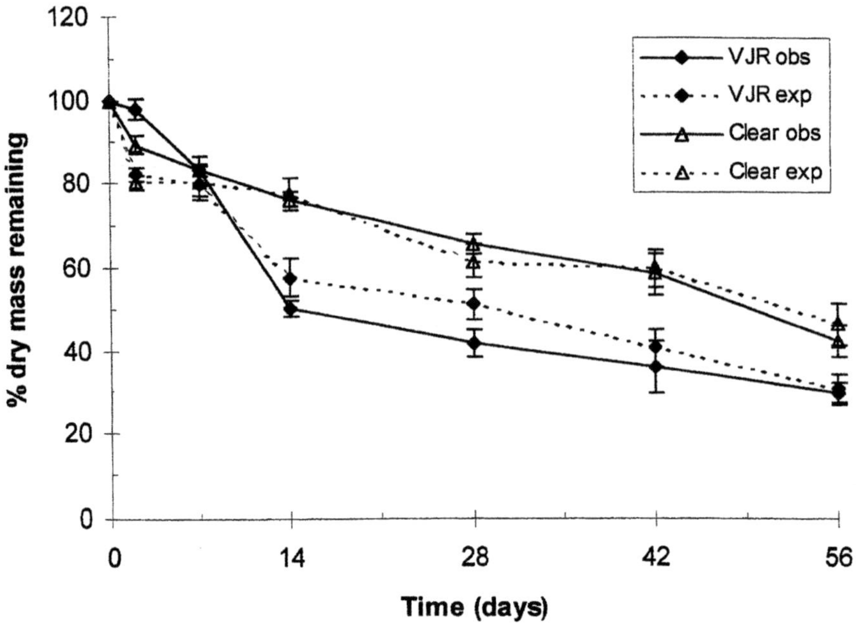

Figure 4. The mean dry mass remaining after time, as % of initial dry mass, for observed mixed species litter bags and expected mixed species (the combination of all the single species) at the Cleared and VJR sites.

tides per year which corresponds to Watson's (1928) inundation class 3 (flooded by normal high tides). The VJR was flooded by 7% of flood tides per year, equating to inundation class 4 (flooded by spring high tides). The horizontal distance from the shoreline was matched, but not the vertical tidal heights because this was impractical. Clearing of trees leads to loss of canopy cover and an anchored root system, and hence protection from tidal erosion, monsoon rain and desiccation (Alongi et al., 1998). The increased erosion of topsoil at the Cleared site probably caused the site to be inundated more frequently as a result of the management regime. Nevertheless, tidal inundation is just one of a number of factors that play a role in leaf decomposition.

Leaves at the Cleared site were subjected to more frequent flooding than the VJR but they decomposed more slowly. Given the more rapid breakdown of immersed leaves, this is unexpected. Decomposition is facilitated in low salinity and aerobic conditions (Mall et al., 1991). Since the salinity at the Cleared site is lower than at the VJR, other factors must be influencing the slower rate of decomposition. Soil redox potential was more negative at the Cleared site than at the VJR (Table 1), indicating anaerobic conditions, which would retard microbial activity. This agrees with Alongi et al. (1998) who attributed low redox potentials at a cleared site to the absence of extensive below-ground living roots. They also recorded low total organic nitrogen concentrations, compared with the VJR site, a further indication of low heterotrophic activity (Snedaker & Snedaker, 1984).

The Cleared site was very exposed to the sun, due to the lack of canopy cover. The VJR in contrast was well shaded. This results in higher air and soil temperatures at the Cleared site (Table 1) and may be another reason for the difference in rates of decomposition. Higher temperatures might enhance microbial activity, increasing nutrient content of litter and accelerating decomposition rate (Mackey & Smail, 1996). However, the bacteria and fungi required for decomposition may not be able to tolerate such high temperatures and loss of moisture, caused by the lack of tree cover, resulting in slower leaf breakdown in the Cleared site. Robertson (1988) comments that leaching and saprophytic decay are more effective when leaves are not subjected to drying.

The lack of tree cover also reduces the amount of leaf litter on the surface of the mud at the Cleared site (6 leaves m^{-2}), compared to the standing stocks of leaf litter at the VJR (43 leaves m^{-2}) (Table 1). The leaf litter at the Cleared site was from a combination of leaves shed by seedlings and saplings and brought in by the wind and tide. The higher rate of leaf breakdown in the VJR may be linked to the greater abundance of decaying leaves.

Table 3. Decay constants (*K*) from exponential equations and half lives (t_{50}) in days for leaf litter bags. Same letters denote litter bags not significantly different from the controls at the $P < 0.01$ level

Species	Site	Intercept	*K* (d^{-1})	R^2	t_{50} (days)
Rhizophora apiculata	Cleared[a]	93.74	−0.0091	0.635	76
	VJR	94.33	−0.0163	0.884	43
Rhizophora mucronata	Cleared[a]	90.94	−0.0057	0.666	122
	VJR	97.94	−0.0204	0.788	34
Bruguiera parviflora	Cleared[a]	83.85	−0.0057	0.618	122
	VJR	80.92	−0.0099	0.747	70
Sonneratia alba	Cleared	79.02	−0.031	0.663	22
	VJR[b]	71.17	−0.0469	0.702	15
Mixed	Cleared	93.54	−0.0135	0.800	51
	VJR	87.62	−0.0218	0.797	32
Mixed	Air[a]	71.32	−0.0065	0.415	107
Mixed	Water[b]	83.67	−0.0668	0.753	10

Table 4. Mixed species litter bags observed and expected exponential equations and half lives for each site

Site	Intercept	*K* (d^{-1})	R^2 (days)	t_{50}
Cleared observed	93.54	−0.0135	0.800	51
Cleared expected	86.12	−0.0128	0.238	54
VJR observed	87.62	−0.0218	0.797	32
VJR expected	84.54	−0.0231	0.347	30

Sonneratia alba leaves decomposed much more rapidly than those of the other species, *Rhizophora apiculata, R. mucronata* and *Bruguiera parviflora*, at both the Cleared and VJR sites. The different species are exposed to the same environmental conditions at each site, so differences in decomposition rate are due to differences in leaf morphology, texture and composition. Differences in leaf quality are known to affect decomposition rates (Robertson, 1988; Basaguren & Pozo, 1994; Twilley et al., 1997). This may relate to two key factors in leaf composition the tannin concentration and the nutritional value. High tannin concentrations are known to be aversive to detritivores and inhibit microbial activity (Neilson et al., 1986; Steinke et al., 1990). However, there appears to be no published information on the tannin content of *Sonneratia* leaves. Nutritional value is defined as the C:N ratio (low ratio means higher nitrogen concentrations and higher nutritional quality). *Sonneratia* has considerably lower initial C:N ratios than *Rhizophora* and *Bruguiera* (Rao et al., 1994; Wafar et al., 1997) and is therefore of high nutritional quality and would be expected to decompose faster (Wieder & Lang, 1982).

Different decay constants are found for different mangrove species, although decay rates vary with temperature and geographical region (Tam et al., 1998). Boonruang (1978) determined a half-life of 40 d for *Rhizophora apiculata* from Thailand similar to the value of 43 d for *R. apiculata* at the VJR in the present study. No other studies give values for *Bruguiera parviflora*. A study by Wafar et al. (1997) in India, recorded half-lives of 289, 347 and 365 d for, respectively, *R. apiculata, R. mucronata* and *S. alba*. Presumably, local environmental factors contributed to decomposition rates much lower than those found in the present study, and explain the lack of difference between *Sonneratia* and the other species.

The much slower decomposition of leaves of *Rhizophora*, the dominant species in Matang, has implications for nutrient recycling and thus ecological functioning of the mangrove ecosystem. Leaf breakdown releases significant amounts of nutrients (Steinke et al., 1993a) and matter (Lu & Lin, 1990) which support tree productivity and the mangrove food web. The longer residence time of *Rhizophora* leaves increases the chance of being washed out into the estuary. This in turn decreases the chance of the nutrients being available to the mangrove plants and detritus consumers. This may then have implications for management and conservation of mangroves when they are to be used as multiple use systems and support coastal crab and prawn fisheries.

Breakdown of mixed species leaf litter was initially slower than predicted, subsequently more rapid. Wardle et al. (1997) also found the same pattern with mixed terrestrial tree species litter bags and emphasise that stage of decomposition is an important determinant of whether positive or negative interactions occur. Overall, however, there was no significant difference in rate of leaf breakdown between the observed results for mixed species litter and that predicted (based on monocultures of the component species) at either site in this study. In the present study, therefore, increasing the diversity of the litter by using the mixed species bags does not increase the rate of decomposition.

This is in contrast to the extensive study carried out by Wardle et al. (1997) on mixing litter from different species. In combinations of more than two species, litter breakdown rates differed from that predicted. Depending on the component species, the effect was either positive or negative: when component species had a high initial nitrogen content mixing enhanced the decomposition rate. With mangrove litter, the decomposition rate may, therefore, be enhanced by *Sonneratia* (with high initial nitrogen concentrations) and retarded by *Rhizophora* (with higher levels of inhibitors such as tannins). The absence of any mixing effect in the present study may, therefore, result from the enhancing effect of one species being offset by the retarding effect of another in the particular species combinations used. However, there may be no effect on decomposition with increased species diversity (Naeem et al., 1994).

An effect of mangrove litter species diversity on decomposition rate has only previously been shown by Mall et al. (1991). They show that litter decomposition rates and soil respiration rates are higher in mixed species litter than in monogeneric species litter. Tietjen & Alongi (1990) suggest that a mixed mangrove species litter is advantageous for the growth and maintenance of nematode populations. However, the study by Gee & Somerfield (1997) showed that under similar conditions of sediment composition, salinity and tidal inundation, mangrove diversity (species of tree or their leaves) has little effect on the diversity and community structure of the meiofauna present in the leaf litter. More information on how functional attributes of species combine in diverse communities will help us to understand ecosystem processes and lead to better conservation and management.

In summary, our study demonstrates: 1. a significant difference ($P < 0.001$) in the rate of decomposition between the Cleared site and the Virgin Jungle Reserve, the VJR decomposing faster; 2. a significant difference ($P < 0.001$) in the rate of decomposition between mangrove leaf species, *Sonneratia alba* decomposing fastest; and 3. a high diversity of leaf species (mixed bags) has no effect on the rate of decomposition of the leaves.

Acknowledgements

The U.K. Natural Environment Research Council (GR9/0288) funded this research. We thank the Malaysian Government, the Forestry Department and in particular Dr Chong Ving Ching for their support in Malaysia. Simon Bryant provided help with data collection and sharing the enjoyment of cleaning and preparing litter bags. Calvin Dytham gave helpful comments on statistical analysis.

References

Aksornkoae, S., 1986. Mangrove ecosystem general background. In: Training Course on Life History of Selected Species of Flora and Fauna in Mangrove Ecosystems. UNDP/UNESCO Regional Project (RAS/86/120): 17–23.

Alongi, D. M., A. Sasekumar, F. Tirendi & P. Dixon, 1998. The influence of stand age on benthic decomposition and recycling of organic matter in managed mangrove forests of Malaysia. J. exp. mar. Biol. Ecol. 225: 197–218.

Ashton, E. C., 1999. Biodiversity in Two Managed Mangrove Forests in Peninsular Malaysia. Doctoral thesis, University of York: 411 pp.

Basaguren, A. & J. Pozo, 1994. Leaf litter processing of alder and eucalyptus in the Aguera stream system (Northern Spain) II Macroinvertebrates associated. Arch. Hydrobiol. 132: 57–68.

Boonruang, P., 1978. The degradation rates of mangrove leaves of *Rhizophora apiculata* (Bl.) and *Avicennia marina* (Forsk.) Vierh. at Phuket Island, Thailand. Research Bulletin No. 26. Phuket Marine Biological Centre, Thailand: 7.

Bunt, J. S., K. G. Boto & G. Boto, 1979. A survey method for estimating potential levels of mangrove forest primary production. Mar. Biol. 52: 123–128.

Camilleri, J. C., 1982. Leaf-litter processing by invertebrates in a mangrove forest in Queensland. Mar. Biol. 114: 139–145.

Camilleri, J. C. & G. Ribi, 1986. Leaching of dissolved organic carbon (DOC) from dead leaves, formation of flakes of DOC and feeding on flakes by crustaceans in mangroves. Mar. Biol. 91: 337–344.

Cintron, G. & Y. S. Schaeffer-Novelli, 1984. Methods for studying mangrove structure. In: Snedaker, S. C. & J. C. Snedaker (eds), The Mangrove Ecosystem: Research Methods. UNESCO, Paris: 91–113.

Cundell, A. M., M. S. Brown, R. Stanford & R. Mitchell, 1979. Microbial degradation of *Rhizophora mangle* leaves immersed in the sea. Estuar. coast. Mar. Sci. 9: 281–286.

English, S., C. Wilkinson & V. Baker (eds), 1994. Survey Manual for Tropical Marine Resources. Australian Institute of Marine Science, Townsville, Australia: 119–196.

Fell, J. W., R. C. Cefalu, I. M. Masters & A. S. Tallman, 1975. Microbial activity in the mangrove (*Rhizophora mangle*) leaf detritus system. In Walsh, G. E., S. C. Snedaker & H. J. Teas (eds), Proc. Int. Symp. Biology and Management of Mangroves, Honolulu, 1974, vol II. University of Florida, Gainesville Florida: 23–42.

Gan, B. K., 1995. A working plan for the Matang mangrove forests Perak (fourth revision). Published by the State Forest Department of Perak Darul Ridzuan, Malaysia: 214.

Gee, J. M. & P. J. Somerfield, 1997. Do mangrove diversity and leaf litter decay promote meiofaunal diversity? J. exp. mar. Biol. Ecol. 218: 13–33.

Golley, F., H. T. Odum & R. F. Wilson, 1962. The structure and metabolism of a Puerto Rico mangrove forest in May. Ecology 43: 9–19.

Gong, W. K. & J. E. Ong, 1995. The use of demographic studies in mangrove silviculture. Hydrobiologia 295: 255–261.

Gong, W. K., J. E. Ong, C. H. Wong & G. Dhanarajan, 1984. Productivity of mangrove trees and its significance in a managed mangrove ecosystem in Malaysia. In Soepadmo, E., A. N. Rao & D. J. Macintosh (eds), Proc. UNESCO As. Symp. Mangr. Env.-Res and Manag. University Malaya, Malaysia: 216–225.

Haron, H. A. H., 1981. A working plan for the second 30-year rotation of the Matang Mangrove Forest Reserve Perak, the first 10-year period 1980–1989. State Forestry Department, Perak: 115.

Hooper, D. U. & P. M. Vitousek, 1998. Effects of plant composition and diversity on nutrient cycling. Ecol. Mono 68: 121–149.

Jolliffe, P. A., 1997. Are mixed populations of plant species more productive than pure stands? Oikos 80: 595–606.

Lee, S. Y., 1995. Mangrove outwelling: a review. Hydrobiologia 295: 203–212.

Lu, C. & P. Lin, 1990. Studies on litter fall and decomposition of *Bruguiera sexangula* (Lour.) Poir, community on Hainan Island, China. Bull. Mar. Sci. 47: 139–148.

Mackey, A. P. & G. Smail, 1996. The decomposition of mangrove litter in a subtropical mangrove forest. Hydrobiologia 332: 93–98.

Mall, L. P., V. P. Singh & A. Garge, 1991. Study of biomass, litterfall, litter decomposition and soil respiration in monogeneric and mixed mangrove forests of Andaman Islands. Trop. Ecol. 32: 144–152.

McNaughton, S. J., 1993. Biodiversity and function of grazing ecosystems. In Schulze, E. D. & H. A. Mooney (eds), Biodiversity and Ecosystem Function. Springer, Berlin: 361–383.

Naeem, S. & S. Li, 1997. Biodiversity enhances ecosystem reliability. Nature 390: 507–509.

Naeem, S., L. Thompson, S. Lawler, J. H. Lawton & R. M. Woodfin, 1994. Declining biodiversity can alter the performance of ecosystems. Nature 368: 734–737.

Neilson, M. J. & G. N. Richards, 1989. Chemical composition of degrading mangrove leaf litter and changes produced after consumption by mangrove crab *Neosarmatium smithii* (Crustacea: Decapoda: Sesarmidae). J. Chem. Ecol. 15: 1267–1283.

Neilson, M. J., R. L. Giddins & G. N. Richards, 1986. Effects of tannins on the palatibility of mangrove leaves in the tropical sesarmid crab *Neosarmatium smithii* (Crustacea: Decapoda: Sesarmidae). Mar. Ecol. Prog. Ser. 34: 185–186.

Newell, S. Y., J. W. Fell, A. Statzell-Tallman, C. Miller & R. Cefalu, 1984. Carbon and nitrogen dynamics in decomposing leaves of three coastal marine vascular plants of the subtropics. Aquat. Bot. 19: 183–192.

Odum, W. E. & E. J. Heald, 1975. The detritus-based food web of an estuarine mangrove community. In Cronin, L. E. (ed.), Estuarine Research. Academic Press, New York: 265–286.

Ong, J. E., 1995. The ecology of mangrove conservation and management. Hydrobiologia 295: 343–351.

Ong, J. E., W. K. Gong, C. H. Wong & G. Dhanarajan, 1984. Contribution of aquatic productivity in managed mangrove ecosystem in Malaysia. In Soepadmo, E., A. N. Rao & D. J. Macintosh (eds), Proc. UNESCO As. Symp. Mangr. Env.-Res and Manag. University Malaya, Malaysia: 209–215.

Poovachiranon, S., K. Boto & N. Duke, 1986. Food preference studies and ingestion rate measurements of the mangrove amphipod *Parhyale hawaiensis*. J. exp. mar. Biol. Ecol. 98: 129–140.

Rao, R. G., A. F. Woitchik, L. Goeyens, A. Vanriet, J. Kazungu & F. Dehairs, 1994. Carbon, nitrogen contents and stable carbon-isotope abundance in mangrove leaves from an East-African coastal lagoon (Kenya). Aquat. Bot. 47: 175–183.

Rice, D. L. & K. R. Tenore, 1981. Dynamics of carbon and nitrogen during the decomposition of detritus derived from estuarine macrophytes. Estuar. coast. shelf Sci. 13: 681–690.

Robertson, A. I., 1988. Decomposition of mangrove leaf litter in tropical Australia. J. exp. mar. Biol. Ecol. 116: 235–247.

Robertson, A. I., D. M. Alongi & K. G. Boto, 1992. Food chains and carbon fluxes. In Robertson, A. I. & D. M. Alongi (eds), Tropical Mangrove Ecosystems. American Geophysical Union, Washington, DC: 293–326.

Schleyer, M. H., 1986. Decomposition in estuarine ecosystems. J. Limnol. Soc. South Afr. 12: 90–98.

Snedaker, S. C. & J. G. Snedaker (eds), 1984. The Mangrove Ecosystem: Research Methods. Monographs on oceanographic methodology 8. UNESCO, Paris: 251.

Spalding, M. D., F. Blasco & C. D. Field (eds), 1997. World Mangrove Atlas. The International Society for Mangrove Ecosystems, Okinawa, Japan: 178.

Steinke, T. D., A. D. Barnabas & R. Samuru, 1990. Structural changes and associated microbial activity accompanying decomposition of mangrove leaves in Mgeni Estuary. S. Afr. J. Bot. 56: 39–48.

Steinke, T. D., G. Naidoo & L. M. Charles, 1983. Degradation of mangrove leaf and stem tissues in situ in mgeni Estuary, South Africa. In Teas, H. J. (ed.). Tasks for Vegetation Science 8. Dr W. Junk Publishers, The Hague: 141–149.

Steinke, T. D., A. J. Holland & Y. Singh, 1993a. Leaching losses during decomposition of mangrove leaf litter. S. Afr. J. Bot. 59: 21–25.

Steinke, T. D., A. Rajh & A. J. Holland, 1993b. The feeding behaviour of the red mangrove crab *Sesarma meinerti* De Man, 1887 (Crustacea: Decapoda: Grapsidae) and its effect on the degradation of mangrove leaf litter. S. Afr. J. Mar. Sci. 13: 151–160.

Stewart, B. A. & B. R. Davies, 1989. The influence of different litterbag designs on the breakdown of leaf material in a small mountain stream. Hydrobiologia 183: 173–177.

Tam, N. F. Y., L. L. P. Vrijmoed & Y. S. Wong, 1990. Nutrient dynamics associated with leaf decomposition in a small subtropical mangrove community in Hong Kong. Bull. Mar. Sci. 47: 68–78.

Tam, N. F. Y., Y. S. Wong, C. Y. Lan & L. N. Wang, 1998. Litter production and decomposition in a subtropical mangrove swamp receiving wastewater. J. exp. mar. Biol. Ecol. 226: 1–18.

Tietjen, J. H. & D. M. Alongi, 1990. Population growth and effects of nematodes on nutrient regeneration and bacteria associated with mangrove detritus from northeastern Queensland (Australia). Mar. Ecol. Prog. Ser. 68: 169–179.

Tilman, D., D. Wedin & J. Knops, 1996. Productivity and sustainability influenced by biodiversity in grassland ecosystems. Nature 379: 718–720.

Twilley, R. R., M. Pozo, V. H. Garcia, V.H. Rivera-Monroy, R. Zambrano & A. Bodero, 1997. Litter dynamics in riverine mangrove forests in the Guayas River estuary, Ecuador. Oecologia 111: 109–122.

Van der Valk, A. G. & P. M. Attiwill, 1984. Decomposition of leaf and root litter of *Avicennia marina* at Westernport Bay, Victoria, Australia. Aquat. Bot. 18: 205–221.

Wafar, S., A. G. Untawale & M. Wafar, 1997. Litterfall and energy flux in a mangrove ecosystem. Estuar. coast. shelf Sci. 44: 111–124.

Wardle, D. A., K. I. Bonner & K. S. Nicholson, 1997. Biodiversity and plant litter: experimental evidence which does not support the view that enhanced species richness improves ecosystem function. Oikos 79: 247–258.

Watson, J. G., 1928. Mangrove forests of the Malay Peninsula. Malayan Forest Records No. 6: 1–275.

Wieder, R. K. & G. E. Lang, 1982. A critique of the analytical methods used in examining decomposition data obtained from litter bags. Ecology 63: 1636–1642.

Hydrobiologia **413:** 89–93, 1999.
R.S. Dodd (ed.), Diversity and Function in Mangrove Ecosystems.

Insect herbivory on mangrove leaves in Guadeloupe: effects on biomass and mineral content

E. Saur[1], D. Imbert[1], J. Etienne[2] & D. Mian[1]
[1]*Université des Antilles et de la Guyane, Biologie et physiologie végétales, 97159, Pointe-à-Pitre, CEDEX, Guadeloupe, Antilles françaises*
[2]*Institut National de la Recherche Agronomique, CR Antilles, B.P. 515, 97165, Pointe-à-Pitre, CEDEX, Guadeloupe, Antilles françaises*

Key words: mangrove ecosystems, defoliation, mineral cycling, *Avicennia germinans*, *Rhizophora mangle*, *Junonia evarete*, *Hyblaea purea*, French West Indies

Abstract

Extensive defoliation of *Avicennia germinans* by *Junonia evarete* (Stoll) (*Nymphalidae*) and *Hyblaea purea* (Cramer) (*Hyblaeidae*) was noted following hurricanes Hugo (1989), Luis & Marilyn (1995). The proposed work is about evaluating the direct loss of organic matter and minerals by insect grazing during an inter-hurricane period. The physiological response of the plant was assessed on perforated leaves over a 3 week period. *A. germinans* in two contrasted sites exhibited a leaf area loss close to 4.5% which corresponds to 50 kg D.W. / ha. In a third site, the low consumption rate of about 0.83% could be related to a drop in the insect population following an earlier outbreack and full tree defoliation. Damage in *R. mangle* was exceptionally low (0.2%). Bulk minerals trapped in the folivore pathway was about 1.2 kg N and 60 g P / ha in *A. germinans* stands. Leaf grazing brings a significant accumulation of organic matter and minerals by importing compounds from other tissues of the plant. This reflects a sudden change in leaf metabolism and we may hypothesise scarring reaction and/or secondary compound accumulation. In any case, leaf grazing demands additional energy and minerals to maintain partially grazed leaves. Moreover, such leaves are susceptible to greater loss in the event of repeated insect damage, due to higher concentrations. From these results, it appears that physiological effects of insect grazing have to be considered not only as direct loss in living tissues but also as deep changes in internal fluxes of the carbon and minerals.

Introduction

The major impact of mangrove litterfall on detritivore food-chains has been stressed by numerous authors for tropical tidally influenced habitats (Lugo & Snedaker, 1974). The subsequent mineral cycling has a strong influence on mineral availability for plants and lateral fluxes to neighbouring habitats. Folivores may influence leaf litter quality and regulate nutrient cycling pathways (Choudhury, 1988). Proportion of leaf area lost to herbivory vary widely depending on mangrove species and environmental conditions (Johnstone, 1981; Robertson & Duke, 1987; Robertson, 1991; Anderson & Duke, 1995; Feller, 1995; Stowe, 1995).

The mangrove of Guadeloupe extends over 3000 ha around the Grand Cul-de-Sac marin Lagoon and was described for litter fall and primary production in Imbert & Portecop (1986) and Imbert & Rollet (1989). Extensive defoliation of *Avicennia germinans* by *Junonia evarete* (Stoll) (*Nymphalidae*) and *Hyblaea purea* (Cramer) (*Hyblaeidae*) was noted following hurricanes Hugo (1989), Luis & Marilyn (1995) (Bouchon et al., 1981, authors data). The work reported here aims to estimate consumption of leaves by herbivores in mixed mangrove stands. The effects of the tree species, forest structure and recent defoliation (following hurricane Luis) were investigated. The second goal of this work was to check for physiological responses of leaves to grazing in terms of carbon and mineral stocks. Internal cycling of carbon and minerals is known to be largely influenced by environmental factors and to play a prominent role in long-term mineral allocation (Nambiar & Fife, 1991). Mineral response to mechanical stress, such as leaf grazing, has not been established previously, and understanding this would contribute to our understanding of tree and stand mineral budgets.

Materials and methods

Study sites

Figure 1 depicts the location of the surveyed sites located around the Grand Cul-de-Sac Marin Lagoon, Guadeloupe (61° W–16° N), delimited by a 30 km long barrier reef in the north. Rainfall averages 1800 mm / year, with a distinct dry season occurring from January to March. The whole area which had been severely damaged by hurricane Hugo (1989) was slightly affected by hurricane Luis (1995), 10–15% of the canopy being defoliated (authors data). Sites 1 and 2, located south of the Canal Perrin, are part of a permanent plot system (Imbert & Portecop, 1986; Imbert & Rollet, 1989). The three sites are representative of the three vegetation belts which have been described elsewhere regarding to their ecology (Imbert, 1985) and the impact of hurricane Hugo on forest structure (Imbert et al., 1996).

Site 1 – mixed high forest belt – is presently dominated by 14 m tall *Avicennia germinans* trees which survived hurricane Hugo. The understory is mixed *A. germinans/Rhizophora mangle* post-hurricane regrowth, 6–8 m tall (6-yr old). The soil is peaty with an average salinity of 45 g / l.

Site 2 – mixed dwarf belt – is a dwarf stand dominated by *A. germinans* (6.5 m tall) with *R. mangle* in the understory (3 m tall). This stand was less affected by hurricane Hugo than the two others. It is located just behind the *R. mangle* fringe belt. The soil is a fibrous peat averaging 52 g / l in salinity.

Site 3 – black mangrove outer belt – is a pure *A. germinans* stand (10 m tall), located North of the Rivière à Goyave and growing on clay deposits with an average salinity of 56 g /l. Massive defoliation occurred in November 1995 from *Junonia evarete* and *Hyblaea purea* swarm following the passages of hurricanes Luis and Marilyn. At the time of the sampling, the top of most twigs and small branches was dead while new leaves densely covered main branches and still alive twings.

Herbivory measurements: biomass and mineral content

Consumption of leaf area was estimated in May 1996 by selective sampling of 10 *A. germinans* in all sites and 10 *R. mangle* in Site 1. Sample trees were canopy trees, selected on the basis of their diameters at breast height (DBH), so that all the range of DBH sizes was represented and randomly distributed over a 1 ha area. Leaf sampling was performed by randomly collecting 10 leaves from five branches distributed along the whole stem, using a pruning hook. Grazed leaf area was determined manually using a 1 mm transparent grid over a photocopy. In cases where there was damage to the leaf edge, the outer margin was extrapolated graphically. The original leaf area was calculated from the leaf length (L) and the leaf width (W) applying regression against area (A) established from 150 non-damaged fresh leaves using a laser CI 203 area meter (CID Inc.). The following, significant relationship was established: $A = 0.6759\ L * W$ ($r^2 = 0.973$). Consumption rate was calculated from measured grazed area and estimated original leaf area. Dried samples were used for dry weight, ash weight and determinations of N-P contents.

Experimental design and measurement of biomass and mineral content response to leaf grazing

Experimental design was enhanced for analyzing the physiological response in leaf organic matter and mineral content to grazing stress. At site 1, *A. germinans* leaf grazing was simulated by two perforations (2.3 cm^2 in area) into a fully developed, non damaged leaf which was opposite to an intact one considered as the control. The treatment was randomly replicated 75 times over five trees and 25 pairs of leaves were collected at time 0, 0+7 days and 0+21 days. Area, dry weight and ash content were determined individually for each leaf. Organic matter was calculated as the difference between dry weight and ash weight, and mineral content was considered as the ash weight both expressed per leaf area unit. The relative gains in organic matter and minerals in perforated leaf were calculated as a percentage of the control leave (opposite and not perforated).

Statistics

One-factor analysis of variance was used to test the statistical effects of sites and treatments, then means comparison was conducted by Scheffe *F*-test.

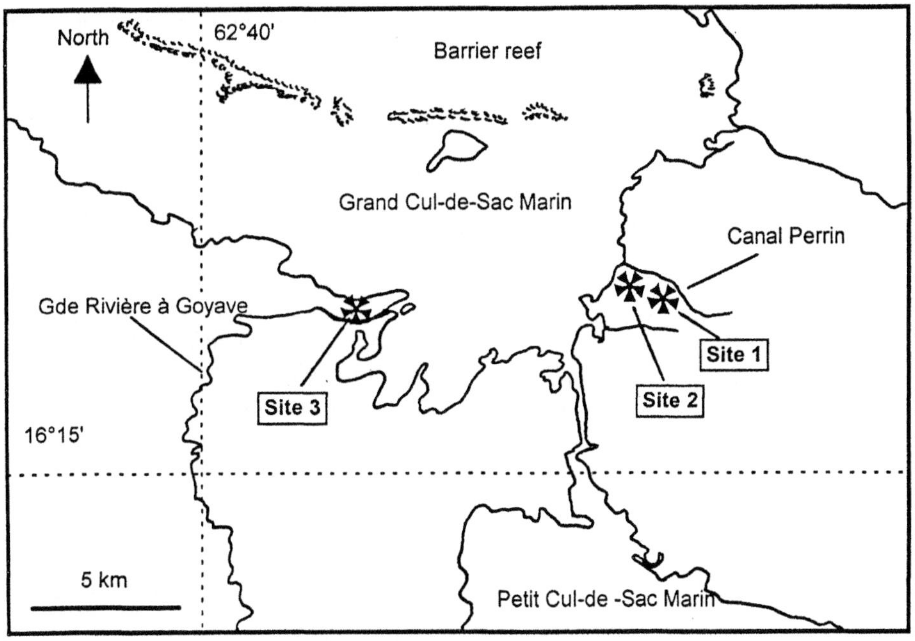

Figure 1. Map of coastal Guadeloupe with sampled sites.

Results

Biomass and mineral content loss from herbivory

Undamaged *A. germinans* leaves were the same size in all sites (Table 1). These results confirm that the Site 3 trees had expended full size leaves by six months after complete defoliation. The percent of individual leaf area damaged was about 4.5% in Sites 1 and 2, and was significantly lower (0.83%) in Site 3. *R. mangle* leaves were nearly without damage considering, that the percentage was close to 0.2%. The highest levels of nitrogen and phosphorus were found in the leaves collected from Site 3 (Table 2). From the mineral content, we evaluated the mineral mass loss by foliovory to average values of 1020 N and 50 P μg / g leaf (D.W.) in site 1 and 2 decreasing to 250 N and 10 P μg / g leaf (D.W.) in Site 3.

Table 1. Insect herbivory measured as leaf area

Species	*A. germinans*			*R. mangle*
Site	I	II	III	I
Full leaf area (cm^2)	28.88 a	27.03 a	26.47 a	39.60 b
Leaf area damaged (cm^2)	1.26 c	0.90 c	0.20 b	0.08 a
Percent of individual leaf area damaged	4.14 c	4.94 c	0.83 b	0.22 a

Within each line, means followed by different letters differ significantly at a confidence level of 5% (Scheffe *F*-test).

Table 2. Nitrogen and phosphorus concentrations in undamaged leaves of *A. germinans* and quantities lost by herbivory

Site	I	II	III
Nitrogen concentration (% D.W.)	2.31 a	2.21 a	3.10 b
Nitrogen lost by folivory (μg / g leaf D.W.)	958 b	1091 b	257 a
Phosphorus concentration (% D.W.)	0.11 a	0.10 a	0.14 b
Phosphorus lost by folivory (μg / g leaf D.W.)	46 b	51 b	11 a

Within each line, means followed by different letters differ significantly at a confidence level of 5% (Scheffe *F*-test).

Biomass and mineral content response to leaf grazing

Organic matter in the control leaves remained constant over three weeks (12 mg D.W. / cm^2). Mineral content decreased slightly from 2.14 to 1.90 mg / cm^2 over the same period of time. Physiological response to grazing stress was measured as the ratio of organic matter and mineral content of perforated *vs.* intact leaves (Figure 2). Results showed a significant increase in both organic matter and minerals during the first week; then the relative gain remained constant over the duration

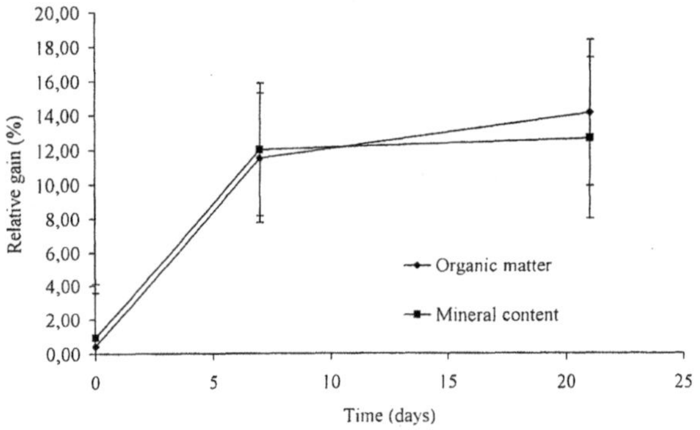

Figure 2. Response of organic and mineral contents of leaves to perforation. The relative gains in perforated leaf were calculated as a percentage of the control leave (opposite and not perforated). *Vertical bars represent the standard deviations.*

of the experiment. Such a pattern clearly indicates a net influx of organic and mineral matter in response to the artificial leaf perforations.

Discussion and conclusions

Herbivory on *A. germinans* and *R. mangle* leaves were investigated in the mangrove of Guadeloupe in a steady state situation avoiding identified insect swarm and massive defoliation. For the comparison of leaf area loss, damage was calculated at each site as the mean percentage of leaf area missing from all leaves of a species sample not corrected for missing leaves which appeared to be negligible and was classically excluded in *Rhizophora* and *Avicennia* herbivory studies (Robertson & Duke, 1987; Farnworth & Ellison, 1991).

Folivory on *A. germinans* leaves showed no significant differences between site 1 and 2, although these two stands exhibited contrasted ecological conditions and differed both in forest structure and primary production (Imbert & Rollet, 1989). Considering the total leaf biomass of 1200 kg D.W. / ha in site 1 found by Imbert & Rollet (1989), the background leaf area loss close to 4.5% corresponds to 54 kg D.W. of leaves / ha. The very low consumption rate in site 3 (0.83%) could be related to a drop in the insect population following the earlier outbreak and full tree defoliation. Spatial variation in leaf damage appears to be very significant depending on insect population dynamics as recorded by Farnsworth & Ellison (1991) on *R. mangle* foliovory ranging from 4 to 25% area and *A. germinans* from 8 to 36%. *A. germinans* is mainly damaged by *J. evarete* known to be specialised to *Verbenaceae* and which is abundant as flying adult at the experimental sites.

Damage to *R. mangle* is exceptionally low in Guadeloupe from 0.2% in this study to 1.2% reported in Imbert & Rollet (1989). We hypothesize a lack of insects grazing on *Rhizophora* in Guadeloupe, as *Rhizophora*'s grazers appeared to be very specific (Farnworth & Ellison, 1991). Leaf chemical composition is probably a major influence on herbivore activity and *Rhizophoraceae* contain high concentrations of soluble tannin which would decrease palatability; furthermore, physical toughness of leaves may also have a role in controlling herbivore activity of the same species (Roberston & Duke, 1987). Turner (1995) found lower tannin content in *Avicennia* compare to *Rhizophora* and generally higher content in young leaves.

Mineral concentrations in *A. germinans* leaves are very similar in Sites 1 and 2 and significantly higher in Site 3. The young leaves which expanded after defoliation (Site 3) have high concentration in nitrogen and phosphorus prior to retranslocation as established in a wide variety of perennial broadleaf trees (Nambiar & Fife, 1991; Saur et al., 1997). In the same manner, defoliation brought a large amount of minerals to the soil cycling pathway momentarily available for trees. We do not discuss here the relationship between levels of leaf herbivory and nitrogen concentration because high nitrogen could reflect high toxicity as well as high palatability depending on the chemistry of N compounds. Such inconsistency in the response of herbivores to nitrogen concentration is illustrated by several studies on mangrove forest (Johnstone, 1981; Farnworth & Ellison, 1991; Feller, 1995).

Bulk minerals trapping in the foliovore pathway are about 1.2 kg N and 60 g P / ha, as estimated on site 1 and 2, and represent very small quantities in the biogeochemical cycle of a mangrove forest which involve tens of kg N and P per ha and year.

The next goal of the study was to point out physiological response to leaf grazing by an experimental design on *A. germinans*. Considering control leaves, results showed a decrease of mineral content over the 3 weeks corresponding to a net efflux of minerals through phloem tissue. The remobilisation of foliar nutrients is an important source of supply to achieve tree growth and constitutes a large contribution to the internal cycling of most of the evergreen trees (Nambiar & Fife, 1991). From this classical pattern, leaf grazing leads to a significant accumulation of organic matter and minerals (Figure 2) from importing compounds from other tissues of the plant. It represents a sudden change in leaf metabolism and we may hypothesize scarring reaction or/and secondary compound accumulation. In any case, leaf grazing costs additional energy and minerals to maintain partially grazed leaves. Moreover, such leaves are susceptible to greater loss in case of repeated insect damage, due to higher concentration. From these results, it appears now that physiological effects of insect grazing have to be considered not only as direct loss in living tissues but also as deep changes in internal fluxes of carbon and minerals.

Light mechanical stress has been tested in this work in relation to leaf damage observed in the field; physiological response to heavy stress (more than 50% of leaf area) is still unknown and should drive two opposite fluxes (retranslocation) as established for light stresses inducing senescence (Authors data).

Acknowledgements

We gratefully acknowledge technical assistance of Emile Timodent in the field and the laboratory. We thank K. Lacoste for improving the English.

References

Anderson, C. & S. Y. Lee, 1995. Defoliation of the mangrove *Avicennia marina* in Hong-Kong: cause and consequences. Biotropica 27: 218–226.

Bouchon, C., Y. Bouchon-Navaro, D. Imbert & M. Louis, 1991. Effets de l'ouragan Hugo sur les communautés côtières de Guadeloupe (Antilles françaises). Annales de l'Institut océanographique 67: 5–33.

Choudhury, D., 1988. Herbivore induced changes in leaf litter resource quality: a neglected aspect of herbivory in ecosystems nutrient dynamics. Oikos 51: 389–393.

Farnsworth, E. J. & A. M. Ellison, 1991. Patterns of herbivory in Belizean swamps. Biotropica 24: 555–567.

Feller, I. C., 1995. Effects of nutrient enrichment on growth and herbivory of dwarf red mangrove (*Rhizophora mangle*). Ecol. Monogr. 65: 477–505.

Imbert, D., 1985. Organisation spatio-temporelle des communautés végétales dans la mangrove du Grand Cul-de-sac Marin, Guadeloupe. Phd Thesis, Université de Montpellier, France: 185 pp.

Imbert, D. & J. Portecop, 1986. Etude de la production de litière dans la mangrove de Guadeloupe (Antilles françaises). Acta Oecol. Oecol. Plantarum 7: 379–396.

Imbert, D. & B. Rollet, 1989. Phytomasse aérienne et production primaire dans la mangrove du grand Cul-de-Sac Marin (Guadeloupe, Antilles françaises). Bull. Ecol. 20: 27–39.

Imbert, D., P. Labbé & A. Rousteau, 1996. Hurricane damage and forest structure in Guadeloupe, French West Indies. J. Trop. Ecol. 12: 663–680.

Johnstone, I. M., 1981. Consumption of leaves by herbivores in mixed mangrove stands. Biotropica 13: 252–259.

Lugo, A. E. & S. C. Snedaker, 1974. The ecology of mangrove. Ann. Rev. Ecol. Syst. 5: 39–64.

Nambiar, E. K. S. & D. N. Fife, 1991. Nutrient retranslocation in temperate conifers. Tree Physiol. 9: 185–207.

Robertson, A. I., 1991. Plant-animal interactions and the structure and function of mangrove forest ecosystems. Aust. J. Ecol. 16: 433–443.

Robertson, A. I. & N. C. Duke, 1987. Insect herbivory on mangrove leaves in North Queensland. Aust. J. Ecol. 12: 1–7.

Saur E., D. N. Fife & E. K. S. Nambiar, 1997. Seasonal changes in accumulation and retranslocation of nutrients in *Casuarina glauca* cladodes. In T. Ando et al. (eds), Plant Nutrition – for Sustainable Food Production and Environment. Kluwer Academic Publishers: 171–172.

Stowe, K. A., 1995. Intracrown distribution of herbivore damage on *Laguncularia racemosa* in a tidally influenced riparian habitat. Biotropica 27: 509–512.

Turner, I. M., 1995. Foliar defences and habitat adversity of three woody plant communities in Singapore. Funct. Ecol. 9: 279–284.

Hydrobiologia **413:** 95–102, 1999.
R.S. Dodd (ed.), Diversity and Function in Mangrove Ecosystems.

Mangrove-saltmarsh dynamics on a bay-head delta in the Hawkesbury River estuary, New South Wales, Australia

N. Saintilan[1] & T. R. Hashimoto
[1]*School of Arts and Sciences, Australian Catholic University, P.O. Box 968, North Sydney 2059, Australia*
[2]*Division of Geography, School of Geosciences, University of Sydney, 2006, Australia*

Key words: mangrove, saltmarsh, delta, estuary sea-level, accretion, coastal wetlands, Holocene, succession

Abstract

Mangroves of the species *Avicennia marina* and *Aegiceras corniculatum* occupy a range of intertidal environments in the Hawkesbury River estuary, New South Wales, Australia. Over the previous four decades, these species have, in a number of locations, encroached upon upper-intertidal saltmarsh environments. The authors have investigated the upper stratigraphy of saltmarsh plains to determine whether evidence can be found of prior occupation of these sites by mangroves. On the prograding bayhead deltas, mangrove root systems were found preserved at approximately 0.3 m beneath the current saltmarsh surface indicating that vegetation progression has occurred from mangrove to saltmarsh with the accretion of the flat. The mangrove roots dated from 500 to 1700 years before present, indicating that the current landward transgression of mangroves is contrary to the longer-term successional trend in these environments.

Introduction

One of the predicted outcomes of sea-level rise is the landward translation of the mangrove zone (Woodroffe, 1990). Studies from Florida (Alexander, 1974; Alexander & Crook, 1974) and Australia (Vanderzee, 1988; Saintilan & Williams, 1999) have described the extension of mangroves into terrestrial vegetation in this context, with the suggestion that mangrove transgression may be an indicator of relative sea-level rise.

A growing body of evidence has suggested that a landward incursion of mangrove into saltmarsh is a consistent trend across temperate Southeast Australia and is a potential threat to the extent and biodiversity of saltmarsh (Saintilan & Williams, 1999). This trend has been identified in 22 separate estuaries, including the Ramsar-listed wetlands of Moreton Bay, Queensland (Morton, 1994), Towra Point, New South Wales (Mitchell & Adam, 1989a) and Corner Inlet, Victoria (Vanderzee, 1988). In some situations, saltmarsh loss has exceeded 70% in the previous three decades (Evans, 1997).

Determining the principal causes of this change is difficult, given the number of factors which could be involved. Suggestions have been varied, including the recolonization by mangroves of areas from which they have been cleared (Mitchell & Adam, 1989a), the response of mangroves to eustatic increases in sea-level (Saintilan & Williams, 1999) and the improved conditions for colonization of mangroves in higher intertidal environments in the presence of additional nutrients and sediment following the development of catchments (Wilton, 1998).

It has further been suggested that saltmarsh is a pioneer community and is replaced by mangrove in temporal succession (Mitchell & Adam, 1989b; Morrisey, 1995). This theory of successional change contradicts the more traditional view that in the absence of changes in relative sea-level, mangroves are progressively replaced over time by saltmarsh, as the flat is vertically accreted by tidally-borne sediments (Pidgeon, 1930). Mitchell & Adam (1989b) suggested this theory after stratigraphic studies in the Georges and Hacking Rivers failed to identify the presence of mangrove roots beneath contemporary saltmarsh.

However, the recent discovery of mangrove roots preserved beneath the saltmarsh plains of the Hawkesbury River supports the traditional model of depositionally-driven succession (Saintilan, 1997), at least on prograding fluvial deltas. The present paper presents the stratigraphic position and age of these mangrove root systems and places contemporary changes in the context of longer-term mangrove-saltmarsh dynamics on the Hawkesbury River.

Description of sites studied

The Hawkesbury River occupies a deeply-incised valley of Triassic quartz sandstone (Conaghan, 1980; Herbert, 1983). The estuary formed as the result of the postglacial marine transgression which initially converted the fluvial valley into an estuarine central basin. Infilling of the central basin with fluvial muds was rapid, essentially keeping pace with sea-level rise (Roy, 1984). The present day sea-level was attained at approximately 6400 years B.P. and has since been relatively stable (Thom & Roy, 1983). The progradation of the sandy bay-head delta in the upper estuary over these fluvial muds commenced less than 500 years after the start of this stillstand (Nichol et al., 1995). However, evidence from the lower estuary suggests that some fluvial sand was already being transported into the central basin during the transgression (Roy, 1984). Today, the lower bay-head delta is dominated by subtidal and intertidal point bars, levees and distributary mouth bars, while the upper delta comprises a subaerial alluvial plain traversed by a leveed channel. The channel is tidal to Richmond, 106 km from the coast.

Materials and methods

Survey techniques

The lower bay-head delta is the principal habitat of mangroves and saltmarshes in the Hawkesbury system (Saintilan, 1997), with the most extensive area at the confluence of the Hawkesbury River and Mangrove Creek, 27 km from the Hawkesbury mouth. The saltmarsh flora consists principally of *Juncus krausii*, *Sarcocornia quinquiflora*, *Sporobolus virginicus*, *Somolus repens* and a landward fringe of *Casuarina glauca*.

Mangroves of the species *Avicennia marina* and *Aegiceras corniculatum* form a sympatric association. Contemporary distributions of mangrove and saltmarsh were mapped using the 1994 Sydney air photograph series and ground truthed in January 1997. Comparisons of distributions were then made at a 1: 25000 scale using the 1954 Gosford series (Run 11).

Initial reconnaissance consisted of seven bank-normal transects of four to eight cores to 1.8 m depth at the mouth of the Glenworth Valley, Mangrove Creek and Couranga Point and Gentlemans Halt in the Hawkesbury Valley. The survey revealed decaying mangrove root material beneath saltmarsh within Mangrove Creek and Allens Creek and these sites became the location of more detailed stratigraphic analysis.

Four locations were selected for the dating of preserved mangrove root material. Within Mangrove Creek, two locations were chosen within 100 m of the contemporary mangrove-saltmarsh boundary (Figure 1 Samples 1, 2), with an additional sample collected closer to the landward edge of the saltmarsh (Figure 1 Sample 2A). One sample was collected from Allens Creek (Figure 1, Sample 3) close to the landward edge of the saltmarsh plain. Shell material was extracted for dating from estuarine / deltaic sediment beneath Sample 2 (Mangrove Creek) and Sample 3 (Allens Creek).

Cores were extracted from the sites of samples 1, 2 and 3 (cores H1, H2 and H3, respectively) for more detailed analysis as described below. At these three locations, transects surveys of vegetation and elevation were undertaken from the landward saltmarsh boundary to the seaward mangrove edge, in order to determine the contemporary elevational characteristics of mangrove and saltmarsh within the tributaries. Elevation with respect to the spring tidal level was determined using poles marked with soluble die, a method described by English et al. (1997).

Dating of shells and roots

Shells were washed and submitted to acid etch before standard radiometric dating. Muds containing roots were excavated and roots retrieved individually from the mud matrix. Roots were then given an acid/alkali/acid pretreatment before undergoing standard radiometric dating. The measured age of shell and root material was converted into the conventional C14 age after applying $^{13}C/^{12}C$ corrections.

Core-logging and subsampling

After collection, the cores were split into halves, then visually logged for sediment colour, field texture,

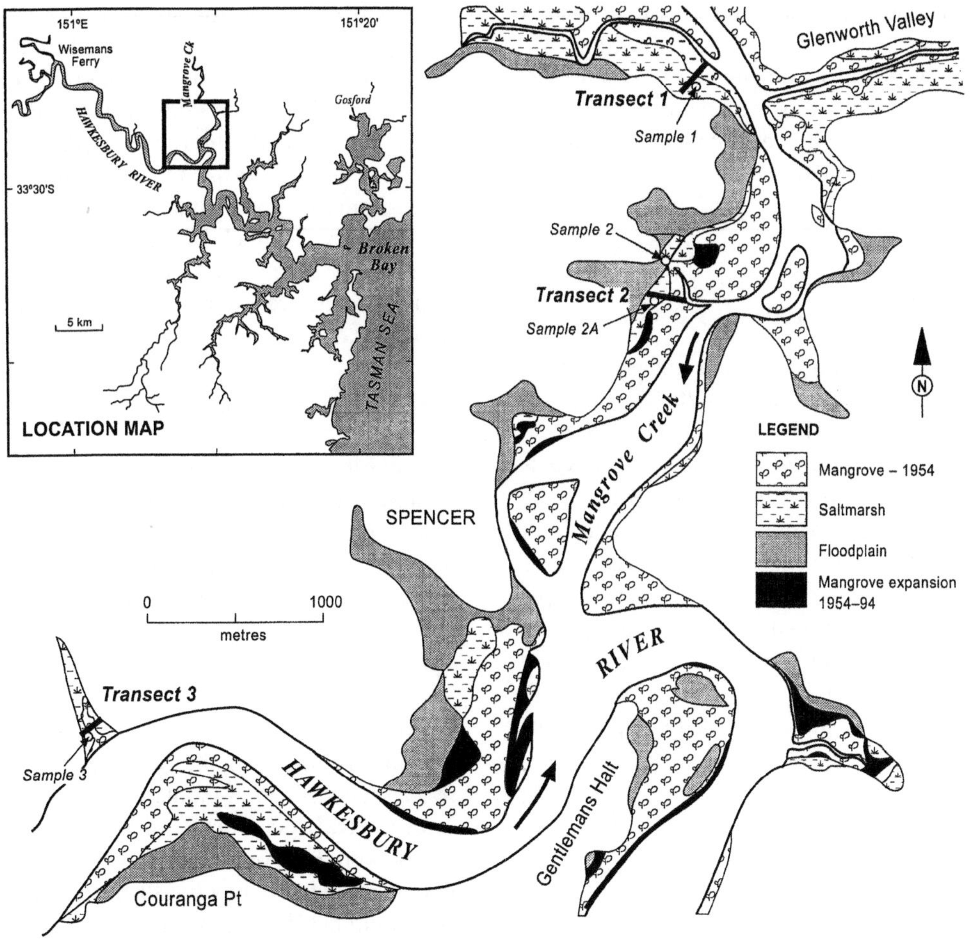

Figure 1. The location of the intertidal bayhead vegetation on the Hawkesbury River estuary. The extent of mangrove transgression in the period 1954–1994 appears in black. Profile and core locations are also shown.

presence of organic matter (particularly *Juncus* and *Avicennia* roots) and shells, and stratigraphic changes. After logging, subsamples were taken from each major stratigraphic unit recognized within the cores for the determination of sediment bulk density, moisture content, organic matter content and gravel / sand / silt / clay ratio.

The bulk density of each subsample was determined by weighing a known volume (20 cm^3) of sediment and applying the formula:

$$\text{Bulk density } (g / cm^3) = \frac{\text{mass (g)}}{\text{volume } (cm^3)}$$

Then, 20 cm^3 sediment samples were oven-dried at 70 °C for 48 h and reweighed, to determine their moisture content, expressed as percentage loss of mass upon drying relative to the wet mass.

To determine the organic content of each subsample, 10 – 16 g of oven-dried sediment was placed in a ceramic crucible and heated to 430 °C for 24 h in an electric furnace. The crucibles were then allowed to cool in a desiccator jar and reweighed. The organic matter content was expressed as percentage loss of mass upon combustion relative to the original dry sediment mass.

The gravel / sand / silt / clay ratio for each subsample was determined through sieving and hydrometer measurements. Wet sieving was carried out on 40 – 60 g of wet sediment disaggregated initially with a dispersant solution (tetra - potassium pyrophosphate) in an ultrasonic bath (30 min). The sediment – dispersant mixture was then passed through a 63 μm mesh stainless steel sieve, whereby the <63 μm fraction and the solution were retained and transferred into a 1000 ml measuring cylinder. The mixture was agitated vigorously for 30 s, then the concentration of suspended solids within the mixture was measured with a hydrometer ($R1$). Another hydrometer measurement ($R2$) was taken 2 h after the first measurement. The relative proportion of clay and silt in the total mud fraction (i.e. clay and silt) was calculated as:

$$\text{Clay}\ (\%) = \frac{R2}{R1} \times 100$$

$$\text{Silt}\ (\%) = \frac{R1 - R2}{R1} \times 100$$

The >63 μm fraction (i.e. sand and gravel) retained in the sieve was oven-dried at 70 °C overnight, then its organic component was removed by hand and combustion at 430 °C for 24 h. Then it was separated into gravel- and sand-fractions with a 2 mm mesh sieve, then reweighed. The ratio of gravel -, sand -, silt - and clay - sized particles was recalculated as a mass percentage of the sediment free of >63 μm organic matter. The organic matter in the <63 μm fraction was not excluded from calculations, due to the likelihood that a significant proportion of this would be allochthonous and, therefore, indicative of the depositional conditions.

Results

Changes in the distribution of mangrove and saltmarsh

A comparison of 1994 and 1954 air photography revealed high rates of incursion of *Avicennia marina* upon saltmarsh (principally communities of *Sarcocornia quinquiflora* and *Sporobolus virginicus)* on intertidal flats adjacent to the Hawkesbury channel near Spencer, Marlows Gully and Couranga Point where 17.5 ha of saltmarsh has been replaced by mangrove (Figure 1). Within Mangrove Creek, the *S. quinquiflora* groundcover of an *Avicennia marina* community has been replaced by *Aegiceras corniculatum* south of transect 2. The extensive plains of *Juncus krausii* in the Glenworth Valley and the upper reaches of the Mangrove Creek estuary do not seem to have been affected, though microcliffing at the mangrove-saltmarsh boundary indicates a slow rate of lateral erosion of the saltmarsh plain.

Location of preserved mangrove roots

Decaying mangrove root systems were found in particularly high concentrations at Allens Creek and the saltmarsh plains on the western bank of Mangrove Creek. While mangrove roots were found beneath saltmarsh further upstream in Mangrove Creek and at the mouth of the Glenworth Valley, they were highly decayed and heavily interpenetrated with the root system of *Juncus krausii.* Older mangrove roots were less commonly found on the point bars of the main Hawkesbury Channel, being absent from Gentlemans Halt and occurring at Couranga Point only in locations currently receiving fresh incursions of juvenile *A.marina.*

Elevation characteristics of mangrove and saltmarsh

Within Mangrove and Allens Creeks, the change from mangrove to saltmarsh occurs at approximately 30 cm below the level of the highest spring tide (Figure 2). Mangroves then occupy the intertidal flat to 60 cm below the level of the highest spring tide.

Sediment characteristics and stratigraphy

At Mangrove Creek (cores H1 and H2) oxidized dark grey-brown mud containing a large amount of *Juncus* rootlets and rhizomes (some live) occurs from the saltmarsh surface to a depth of 25 – 30 cm. At Allens Creek (core H3), oxidized sandy mud and muddy sand extend to 45 cm below the surface. The upper limit of *Avicennia* cable roots occurs at approximately 20 cm below the surface at all locations, within this oxidized zone (Figure 3).

The oxidized sediments are underlain by reduced dark grey mud containing abundant *Avicennia* roots in their upper parts. These roots have been dated at 1700 +/- 90 years BP (Beta 105369) at Allens Creek, 490 +/- 70 years BP (Beta 105370) in core H2 at Mangrove Creek, 1220 +/- 80 years BP (Beta 106995) at a more landward location within the same saltmarsh and 520 +/- 70 years BP (Beta 106994) in core H1, Mangrove Creek. At Allens Creek, the upper part of the reduced zone remains sandy (coarse in places) to 70 cm and becomes more muddy below this level. In all cases, the abundance of *Avicennia* cable roots and fibrous rootlets declines with depth and becomes absent below approximately 170 – 180 cm below the surface. Shell fragments were observed in the lower parts of the range of *Avicennia* roots, below 172 cm in core H2, Mangrove Creek (2660 +/- 70 C14 years BP; Beta 105368) and below 110 cm at Allens Creek (5420 +/- 120 C14 years BP; Beta 105367). At all locations, the mud grades into fine micaceous sandy mud toward the

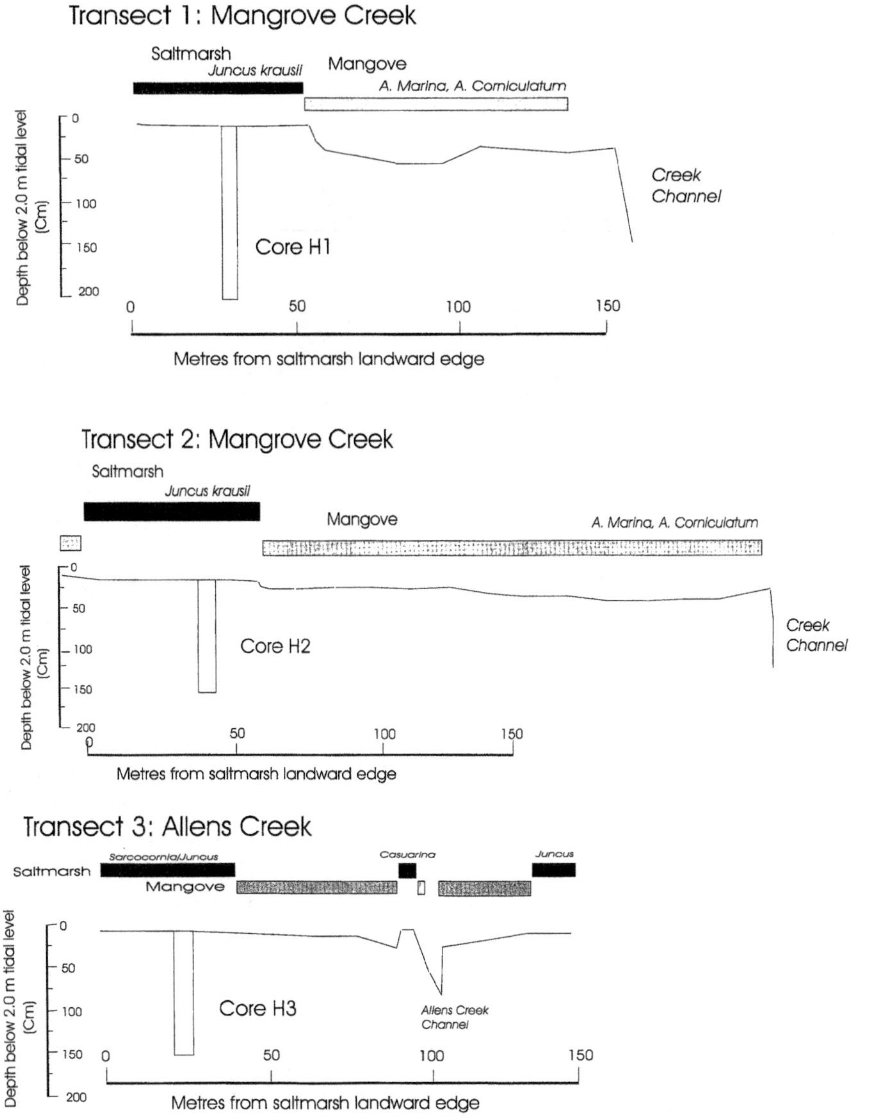

Figure 2. Elevation profiles, vegetation transects and core locations. Refer to Figure 1 for the location of transects.

bottom of the core (below 200 cm at Mangrove Creek, below 150 cm at Allens Creek, Figure 3).

At Mangrove Creek, the highest values of organic matter (10 – 20%) and moisture content (55 – 70%) and the lowest bulk density values (1.2 – 1.4 g / cm^3) occur within the oxidized zone and the upper parts of the reduced *Avicennia*-rich mud. Below this, organic matter / moisture contents decrease, and the bulk density increases downward through the reduced mud into the sandy mud unit at the base of the cores. At Allens Creek, the sandy sediments of the oxidized and the *Avicennia*-rich zones record the highest bulk density (1.83–1.87 g / cm^3) and the lowest moisture content (21 – 33%) values. However, as at Mangrove Creek, the highest organic matter content (12%) is recorded within the *Avicennia*-rich zone.

Discussion

The formation of the intertidal flats in the study area appears to be a consequence of the growth of subaqueous point bars. These point bars were initiated after the expansion of the bay-head delta front through the estuary. As they grew through lateral accretion of sandy fluvial sediments delivered from upstream

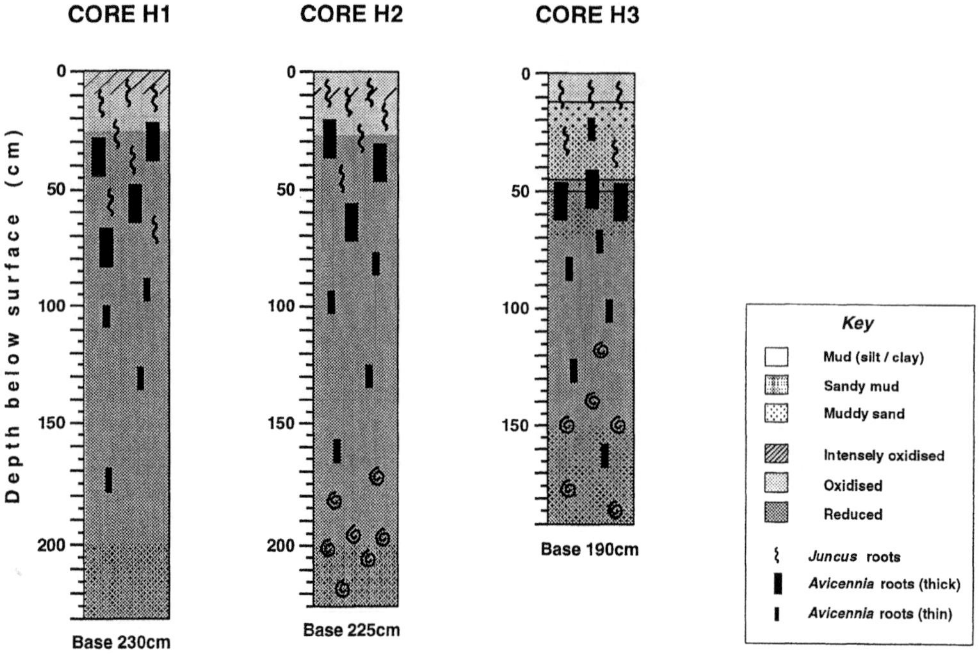

Figure 3. Stratigraphic Sections of cores. Refer to Figure 2 for the location of cores.

during floods, vertical accretion of mud through tidal action began to raise the older (landward) parts of the bar to intertidal elevations, allowing colonization by mangrove initially, and by saltmarsh with continued accretion to upper intertidal levels. Within Mangrove Creek, the age of the preserved mangrove root systems varies from approximately 500 years BP near the seaward edge of the saltmarsh to 1200 years BP near the landward edge, suggesting a gradual replacement of mangrove by saltmarsh landward to seaward in parallel with the growth of the point bar.

The preserved mangrove roots are within the elevation range of contemporary mangrove root systems and are overlain by 20–30 cm of sediment, both clastic and organic. The frequency of tidal inundation of the saltmarsh would decrease rapidly as its surface elevation approaches the upper limit of the intertidal zone. The considerable difference in the age of preserved mangrove roots at Mangrove Creek and Allens Creek, despite the compatible elevations of their current saltmarsh surface, suggests that the rate of vertical accretion slows to a negligible quantity with the decline in frequency of inundation (as proposed by French et al., 1995).

The incursion of mangrove upon saltmarsh was evident throughout the study area, though the rate of encroachment varied between geomorphic settings. Within Mangrove Creek, saltmarsh communities of *Juncus krausii* were subject to slow rates of mangrove encroachment as the seaward edge of the saltmarsh was eroded (Figure 4a).The mechanism is similar to that described for receding saltmarsh in the subsiding coastlines of southeast England and the United States Gulf coast (Guilcher, 1981; Salinas et al., 1986; Phillips, 1986) and has been observed in Corner Inlet, Victoria (Vanderzee, 1988). While the erosion fronts are active and unvegetated, changes were not of a scale detectable from air photo reconnaissance.

Where saltmarsh communities were dominated by *Sarcocornia quinquiflora* and *Sporobolus virginicus*, rates of mangrove encroachment were more rapid. No microcliffing was evident at the mangrove-saltmarsh boundary, and within Mangrove Creek the loss of saltmarsh was primarily due to the replacement of *S. quinquiflora* by the mangrove *Aegiceras corniculatum* (Figure 4c). On the intertidal banks of the main Hawkesbury River channel, the replacement of *S. quinquiflora* is primarily by the grey mangrove *Avicennia marina*. This incursion has been particularly rapid at Couranga Point where mangroves have spread behind a levee bank at the seaward edge of the saltmarsh plain (Figure 4b).

There is little stratigraphic evidence to suggest that post-depositional compaction and subsidence has been

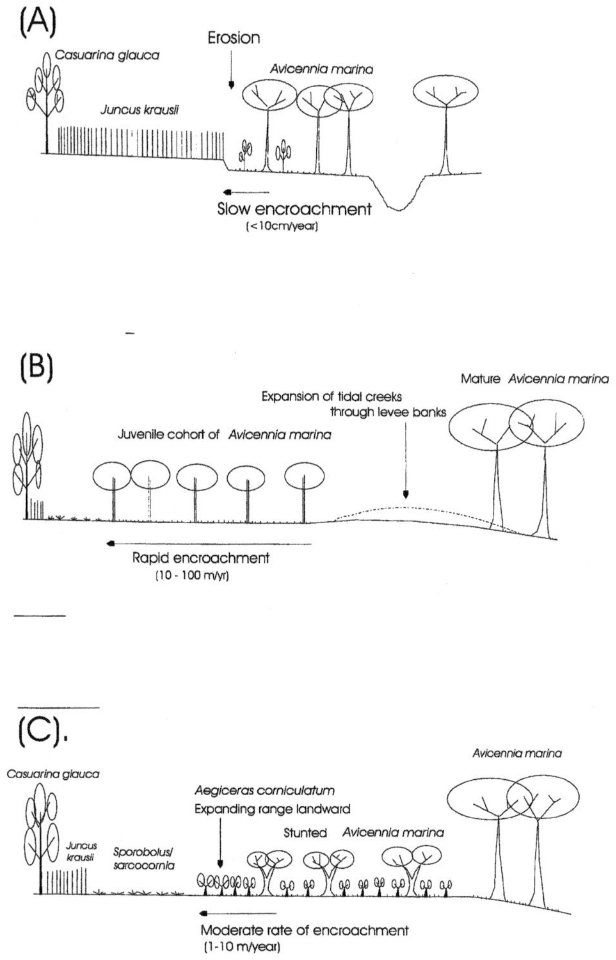

Figure 4. Modes of mangrove encroachment upon saltmarsh in the study area. (a) Recession of saltmarsh along an erosion front. (b) Expansion of the tidal creek network through a frontal levee leading to the rapid colonization of the saltmarsh plain. (c) Expansion of *A. corniculatum* on a gently inclining intertidal slope.

responsible for mangrove transgression. Although the downward increase in sediment bulk density, and a corresponding decrease in the moisture content at Mangrove Creek, and a deeper lower limit of oxidation at Allens Creek seem to suggest sediment compaction over time, the comparable depth of mangrove root occurrence in both areas and the higher upper limit of shell occurrence at Allens Creek, despite the difference in the time of deposition, indicate that compactional subsidence has been minor. In fact, the parallelism in trends suggests that the variations in bulk density and moisture content are controlled by the content of organic matter and sand in the sediment, rather than by any post-depositional changes.

That mangroves are now observed encroaching upon saltmarsh suggests a change in the relationship between sea-level and the rate of sedimentation in the Hawkesbury. It is unlikely that rates of sedimentation and accretion have declined in recent decades. The lateral expansion of mangroves at their seaward edge and a significant narrowing of the channel width (Gardiner, 1995), suggests an increased rate of sediment accumulation over the air photo period. Sea level has risen on the Sydney coastline over the same period, with a particularly high rate of increase in the period 1940–1986 (1.5 mm/yr; Gordon, 1988). Mangroves show their maximal rate of incursion between the air photograph series of 1954 and 1982, which corresponds to the period of maximum sea-level rise allowing for the lag between propagule establishment and growth of a juveniles large enough to be identified in the photographs.

These observations highlight the importance of temporal scale in the prediction of coastal wetland reponses to sea-level change. The average rate of sea-level rise and the average rate of accretion over the previous four decades would not suggest the likelihood of mangrove incursion into saltmarsh, given longer-term trends. However, we propose that a brief period of rapid sea-level rise was sufficient to offset increases in accretion and allow the expansion of mangroves in some, though not all of the geomorphic settings studied.

Acknowledgements

This research was made possible through funding from the Australian Research Council. Peter Johnson is thanked for assistance in the preparation of figures.

References

Alexander, T. R., 1974. Evidence of recent sea level rise derived from ecological studies on Key Largo, Florida. In Gleason, P. J. (ed.), Environments of South Florida: Present and Past. Memoir 2. Miami Geol. Soc. 219–21.

Alexander, T. R. & A. G. Crook, 1974. Recent vegetation change in southern Florida. In Gleason, P. J. (ed.), Environments of South Florida: Present and Past. Memoir 2. Miami Geol. Soc. 61–72.

Conaghan, P. J., 1980. The Hawkesbury sandstone: Gross characteristics and depositional environment. In Herbert, C. & R. Helby (eds), A Guide to the Sydney Basin. Geological Survey of New South Wales, Department of Mineral Resources, Bull. 26: 188–254.

English, S., C. Wilkinson & V. Baker, 1997. Survey manual for tropical marine resources (2nd edn) Australian Institute of Marine Science, Townsville, Queensland.

Evans, M., 1997. Historical distribution of estuarine wetlands at Kurnell Peninsula, Botany Bay – The need and potential for rehabilitation. B.Sc. Honours thesis, University of Wollongong.

French, J., T. Spencer & D. J. Reed, 1995. Geomorphic Responses to sea-level rise: existing evidence and future impacts. Earth Surface Processes and Landforms 20: 1–6.

Gardiner, B., 1993. Hawkesbury-Nepean River Sediment Dynamics Study. Sediment accumulation and sediment cycling in two estuarine depositional environments in the Foul Weather Reach and Spencer areas. Report No. 93/133. Australian Water Technologies, Science and Environment Division. West Ryde, NSW.

Gordon, A. D., 1988. A tentative but tantalizing link between sea-level rise and coastal recession in New South Wales. In Pearman, G. I. (ed.), Greenhouse, Planning for Climatic Change. CSIRO Division of Atmospheric Research, Melbourne.

Guilcher, A., 1981. Shoreline changes in salt marshes and mangrove swamps (mangals) within the past century. In Bird, E. C. F. & K. Koike (eds), Coastal Dynamics and Scientific Sites. 31–53.

Herbert, C., 1983. Geology of the Sydney 1:100 000 sheet 9130, Geological Survey of New South Wales, Department of Mineral Resources: 121–127.

Mitchell, M. L. & P. Adam, 1989a. The decline of saltmarsh in Botany Bay. Wetlands (Aust.) 8: 55–60.

Mitchell, M. L. & P. Adam, 1989b. The relationship between mangrove and saltmarsh communities in the Sydney Region. Wetlands (Aust.) 8: 37–46.

Morrisey, D., 1995. Saltmarshes. Chapter 13. In Underwood, A. J. & M. G. Chapman (eds), Coast. Mar. Ecol. Temperate Australia. UNSW Press, Sydney.

Morton, R. M., 1993. Fluctuations in wetland extent in southern Moreton Bay. In Greenwood, J. G. & N. J. Hall (eds), Future Marine Science in Moreton Bay. School of Marine Science, University of Queensland.

Nichol, S. L., B. A. Zaitlin & B. G. Thom, 1997. The upper Hawkesbury River, New South Wales, Australia: a Holocene example of an estuarine bayhead delta complex. Sedimentology 44: 263–266.

Phillips, J. D., 1986. Coastal submergence and marsh fringe erosion. J. Coast. Res. 2: 427–436.

Pidgeon, I. M., 1940. The ecology of the central coast of New South Wales. III. Types of primary succession. Proc. linn. Soc. N.S.W. 65: 221–249.

Roy, P. S., 1984. Holocene sedimentation histories of estuaries in southeastern Australia. In Hodgkin, E. P. (ed.), Estuarine Environments of the Southern Hemisphere. Department of Conservation and Environment, Western Australia, Bulletin No. 161: 23–59.

Salinas, L. M., R. D. DeLaune & W. H. Patrick, 1986. Changes occurring along rapidly submerging coastal area: Louisiana, USA. J. Coast. Res. 2: 269–284.

Saintilan, N., 1997 Mangroves as successional stages on the Hawkesbury River. Wetlands (Aust.) 16: 99–107.

Saintilan, N., & R. Williams, 1999. Mangrove transgression into saltmarsh environments in New South Wales, Australia. Global Biogeogr. Ecol. Letters. 8: 117–124.

Thom, B. G. & P. S. Roy, 1983. Sea-level change in New South Wales over the past 15 000 years. In Hopley D. (ed.), Australian Sea Levels in the Last 15 000 Years: A Review. Townsville, James Cook University, Monograph Series Occasional Paper No. 3: 64–85.

Vanderzee, M. P., 1988. Changes in saltmarsh vegetation as an early indication of sea-level rise. In Pearman, G. I. (ed.), Greenhouse: Planning for Climatic Change. CSIRO Publications, Melbourne: 147–160.

Wilton, K., 1998. Changes in Mangrove and Saltmarsh Areas in the Sydney Region, with Specific Reference to Careel Bay, Pittwater. M.Sc. Thesis, Macquarie University, Australia.

Woodroffe, C. D., 1990. The impact of sea-level rise on mangrove shorelines. Progr. Physic. Geogr. 14: 483–520.

Hydrobiologia **413:** 103–113, 1999.
R.S. Dodd (ed.), Diversity and Function in Mangrove Ecosystems.

Natural spatial dynamics of mangals through their margins: diagnostic elements

Jean-Michel Lebigre
Institut de Géographie, Université Michel de Montaigne - Bordeaux 3, 11 allée Diogène 33600 Pessac, France

Key words: coastal environment, tidal marshes, mangal, margin, progradation, retreat, Madagascar, Mayotte, Brazil, Gabon

Abstract

The natural spatial dynamics of mangals are part of a larger environment – tidal marshes – which generally also includes highly saline or acidified '*tannes*' (areas without any vegetation or covered with grasses developing at the expense of the mangal) developing at the expense of the mangrove. This dynamic is clearly visible in the pattern of vegetation at the edges of mangals. Their appearance varies considerably depending on where they are situated: facing the sea (seaward front of the mangal) and on the banks of channels or the margins of '*tannes*' and swampy formations (landward front of the mangal). In each case, different forms, revealed by studying the architectural layout of the mangroves, can be compared. These forms, which follow a defined typology, are linked either to the progradation or to the total or partial retreat of the mangal. The identification of these forms by enabling an efficient diagnosis of the mangals dynamics has many possible applications. Notably when it is necessary to restore parts of the mangrove or to develop parts of the coast. This method even if it cannot provide irrefutable conclusions in every case, has the advantage of providing a simpler and more rapid diagnosis than diachronic analysis or aerial photography. More than any other kind of vegetation, mangroves present in many cases remarkable natural dynamics. Particularly because they are subjected to the continual variations in sea level, and to variations in climate and of sedimentation, which are more restrictive than in other ecosystems, they appear condemned to never-ending transformations. In a majority of cases, these changes are perceptible over the period of a decade, or even sometimes annually. This is rare in other ecosystems if one excepts consequences of direct human intervention. Tangible signs of these forces within the coastal areas where mangals grow are apparent in the different types of forest margins which we observed in Gabon, Madagascar (Lebigre, 1990), Brazil (Lebigre & Rebelo Mochel, 1995) and Mayotte (Lebigre, 1997), and which have also been encountered and described by other researchers (notably Baltzer, 1975; Lescure, 1977, 1980; Marius, 1979; Bertrand, 1993). A correct analysis of the mangal margins that have been preserved from direct human intervention make it possible to comprehend the natural phenomena which govern them (sedimentation, sea-level variation) and, thus, costly errors in developing the coastal areas can be avoided.

Characteristics of mangal margins

The study of margins is of primary importance in the study of any forest or semi-forested landscape. Apart from their biological importance, they are the expression of the spatial dynamics of the areas that they delimit. Anthropogenic causes, such as deforestation and/or natural processes are at the origin of these dynamics. The examination and interpretation of these margins may contribute to a better understanding of the recent evolution of regions containing such environments. They can, therefore, be used as indicators of the dynamics of a given environment (Battistini & Lebigre, 1994).

In the case of most forest margins, typology can be based on morphological and floristic criteria. But in the case of mountain forests, swamp forests and mangals, it is necessary to start with a definition of position of the mangal in relation to referential elements generally belonging to the field of topography. More precisely in the case of the mangrove inhabited swamps, this definition involves the relative positions of the sea and of dry land to the mangal. A mangal will present an increasing number of margin types according to the increasing complexity of its swamp structure.

On the **seaward fronts**, the margins give way to the lower part of the strand or low slikke made of barren mud flats and sand accumulations which are below the tide line. These can be likened to the **tidal creek bank** margins which will be studied as an isolated case.

Two different types of **landward fronts** can be defined according to the transitions they imply: either directly with areas of dry land, with '*tannes*' (areas without any vegetation or covered with grasses developing at the expense of the mangal) or with swampy prairies which are a part of the sea swamp like the mangal or the '*tannes*'. The glade margins formed by the '*tannes*' in the middle of the mangals (Figure 2) represent only one particular type of landward front. The same applies to edges facing a coastal line or an island within the marsh.

Natural factors such as sedimentation and the variations of sea levels leave a characteristic mark on each type of margin. However, recent human intervention on the mangal sometimes makes it difficult to discern certain forms of margin. This is notable when clearings have been done or in the case of certain types of development which will not be taken into account in this study.

Progradation fronts

The terms progradation front, colonisation front and pioneer front are used when the mangal has a tendency to colonise an open area which opens out in front of it. The types of margin encountered and their causes vary according to whether they are on a landward or seaward front.

Seaward progradation fronts

Seaward fronts are the most commonly studied. Generally, this kind of progradation is the result of the sedimentary process which elevates the **low slikke**, a part of the strand situated below the average tide level, in a slight negative pulse of the sea level. A number of seaward progradation fronts can be identified.

Diffuse scattering fronts can be identified by an implantation of young pioneer mangroves varying in density and progressively covering the low slikke in the direction of the forest mass. In Madagascar, it is *Avicennia marina* and/or *Sonneratia alba* that are always encountered. In Gabon, this kind of front is typically characterised by *Laguncularia racemosa* occasionally mixed with *Avicennia germinans*. In Brazil, the young mangroves become established in Spartina prairies. *Spartina* are pioneer grasses in the low slikke. This kind of front is generally found in areas where the swell is weak or has lost some of its energy.

If the young mangroves grow densely enough, the margins closely resemble the crescent shaped fronts which are discussed later.

It would be difficult to mistake these fronts for erosion fronts or aborted fronts when making observations with aerial photography. Even more so *in situ*. In the case of an aborted front, one encounters adult or dying mangroves scattered along the margin of the low slikke.

Mangrove copse fronts are easily recognisable by small isolated groups of mangroves spread out across the low slikke. In fact, nothing resembles a mangrove thicket front as much as an erosion front even on the terrain. Lescure (1980) gives a precise description of this kind of front formed by *Laguncularia racemosa* in Guyanna. In Gabon (notably in Soukou-Doukou on the Mondah bay) behind the most recent thickets made up of *Rhizophora racemosa*, larger groups result from an anastomosis and form a transition with the older part of the front identifiable by an irregular line and some patches bare of vegetation. These fronts correspond to sectors where the energy of the swell or tidal currents is stronger than in the previous case.

Characteristic **crescent fronts** are composed of *Rhizophora* spp. They can be observed in Gabon (Plate 1). In aerial photography they appear as lobes of isolated or clumped vegetation. The profile of these fronts is definitely convex. Young mangroves set in under the cover of prop roots and sprout their own prop roots throughout the low slikke. Intermediary types with diffuse scattering fronts are characteristic of margins composed of young *Avicennia marina* or young *Sonneratia alba*, as in Tsigoni (Mayotte Island).

Other types exist. It would be more adapted to call the '*pointes de mangrove*' in Guinea described by Bertrand (1994) '**comb progradation fronts**' (*fronts de progradation en peigne*). They can be found at points where tidal creeks meet, marked by tidal turbulence (the island of Singueya is a good example). The low slikke is veined with channels which are at the origin of '*pointes*' composed of young *Avicennia* with some *Laguncularia* amongst them. Preceding these points, there is a loose implantation of young mangroves which could be called a diffuse scatter fore-front. The

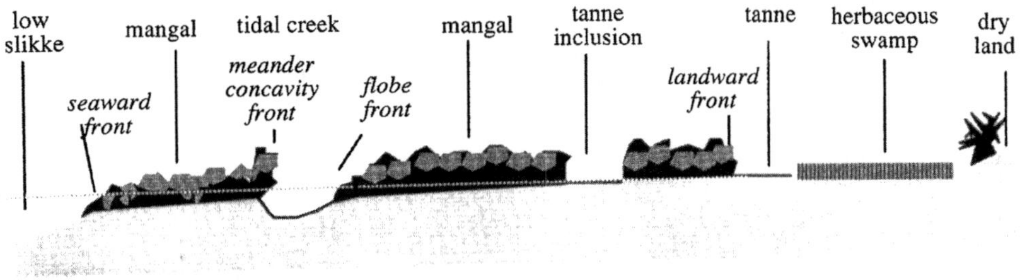

Figure 1. Main mangal margin types by position within the swamp.

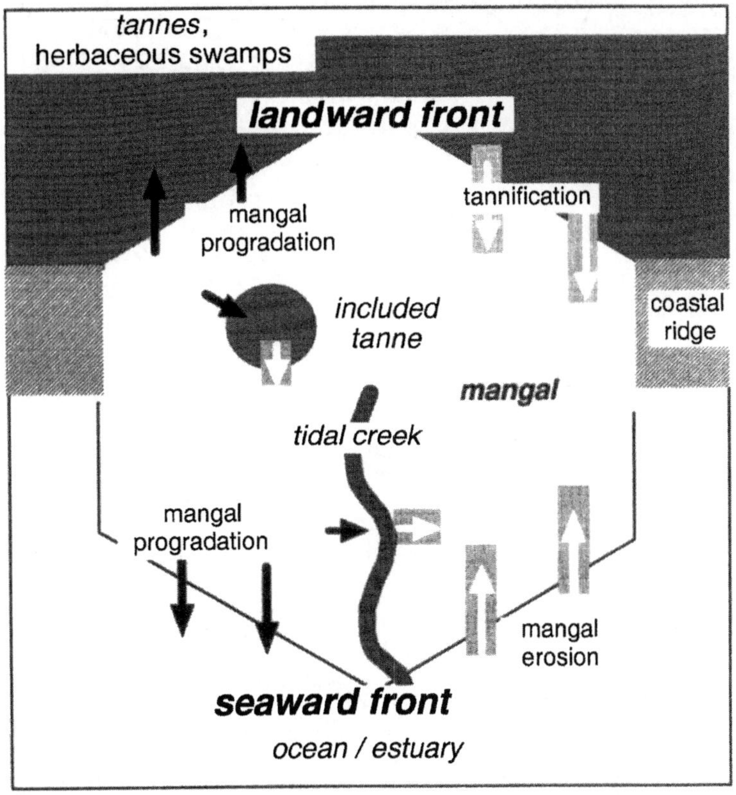

Figure 2. Progradation fronts and retreating fronts by mangal margin position.

author points to haphazardly positioned open patches in populations situated to the back of colonization pockets. However, he is mistaken in saying that a similar type of layout, observed in Gabon, cannot be an erosion front (this is discussed later in this document).

Tidal creek banks

The margins corresponding to the interface between the mangal and tidal channels are less complex. Progradation forms can be observed in the lobes of river bends and along the tidal creeks which tend to pile up (Tsiribihina delta in Madagascar). Here, the equivalent of crescent diffuse scattering fronts on confined spaces. Because of the absence of swell and good sediment levels, the mangroves grow to large sizes rapidly.

Landward progradation fronts

The re-colonization of some *tanne* borders by the mangal might well be a more frequent phenomenon than is usually believed. We have described the banks of the Sao Marcos Bay in Brazil (Lebigre, 1994; Lebigre & Rebelo Mochel, 1995), a type of margin which seems typical of the phenomenon. Indeed, the trans-

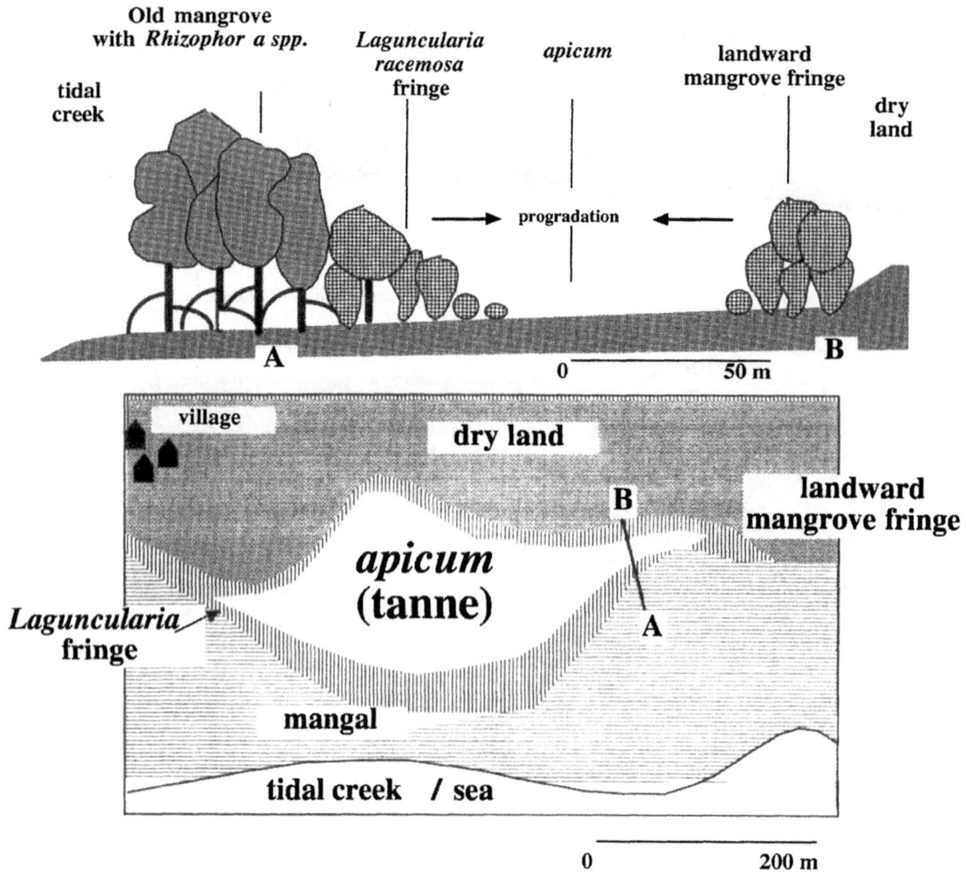

Figure 3. Landward progradation front in the region of Alcantara (Maranhão).

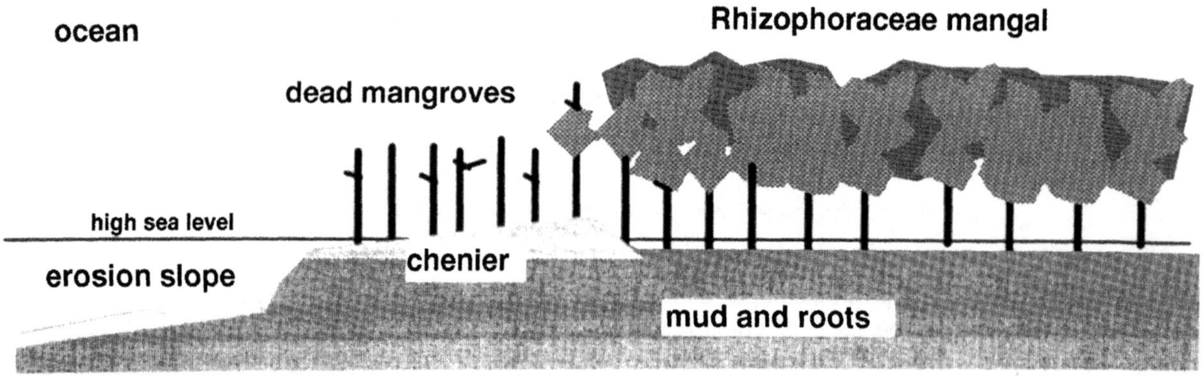

Figure 4. Retreating front outlines (Andopitaly, Tsiribihina delta, Madagascar).

ition between the old clusters of *Rhizophora* and the *tannes* is marked by the presence of a belt of small mangroves – mainly young *Langucularia racemosa* (Plate 2). Around the bigger *tannes*, the belt continues in the direction of the dry land, along what we have called 'the landward fringe of mangroves' (Lebigre, 1990); *Conocarpus erectus* and the *Acrostichum aureum* fern are generally predominant there. Furthermore, groups of small mangroves are frequently observed, scattered between these old clusters of tall trees and the bottom of the hills which constitute the dry land, as if they had replaced former *tannes* (Figure 3). These margins where young individuals proliferate seem to indicate that the *tannes* are in a

Plate 1. Mondah Bay (Gabon): 'crescent front' with *Rhizophora racemosa.*

phase of contraction. This idea is reinforced by the fact that *Laguncularia* has a limited resistance to salt.

This phenomenon is related to that of the internal progradation of the mangal – as described by Lescure (1980) – at the expense of tidal marshes in Guadeloupe, or to Marius's observations in Senegal (1979). Following the important rainfalls of 1978, it can be noted that the herbaceous tanne with *Sesuvium postulacastrum*, resulting from the disappearance of the mangal is taken over by young *Avicennia* in dense stands.

Several hypotheses contradict themselves in attempting to explain this type of phenomenon: an increase in rainfall entailing a decrease in duration or intensity of the dry season in the course of the last decades; a current slight increase of the sea level and a general subsidence which is not compensated by the supply of sediments, an anthropogenic origin of the *tannes* due to the previous destruction of the mangal.

Symptoms of a retreat of the mangal

The retreat of the mangal is apparent in various easily recognisable phenomena encountered on the margins. It entails very different process according to whether we are dealing with landward or seaward fronts.

Seaward retreating fronts and aborted fronts

Seaward retreating fronts generally stem from a retreat of the coastline itself. Most of the time, this last phenomenon results from a particular crisis, limited in time such as a tropical cyclone, a storm or an exceptional tide. Since this type of cataclysm is not always compensated by new supplies of sediments, it sometimes takes a long time for the wounds to heal, and in this case, the vegetation becomes a record of phenomena which are difficult to date.

The main indicators of a retreat are the predominance of senescent mangroves, the loosening of the root system and peculiar outlines.

Characteristic mangrove population

It is generally easy to identify the existence of a retreating front by examining the mangroves which form it. Whereas the progradation front is formed of young, still growing mangroves, the retreating front is generally distinguished by the presence of adult or senescent mangroves, or even of withering individu-

Plate 2. Alcântara in the Maranhão (Brasil): landward progradation front. Young mangroves (*Langucularia racemosa*) seem to be settling over the tanne, called *apicum* in this region.

als. The density of the stand (calculated by number of trees for a given suface area) is consequently quite inferior than in the previous cases. Big isolated trees are even to be found on the strand (Plate 3).

Peat outcrops and loosening of the root system

Peat banks or simple peat outcrops at root level are characteristic of a retreating process. But for this phenomenon to take place, it is vital that the mangrove stand be able to create such peats and this requires long periods for the Rhizophoraeous plants, and more specifically for *Rhizophora* spp. to settle. On the contrary, on grounds where only *Avicennia* spp. or *Sonneratia alba* have been growing, such matter will not be found; but the ground is compacted and contains many rootlets (Plate 4). More frequently, a few decimeters of the mangrove root system are simply loose.

Retreating front outlines

Retreating front outlines are very varied. However, it would be falacious to use this sole criterion, only verifiable on aerial photographs or high resolution spatial images, to highlight mangal retreats. Indeed, there are great similarities with the outlines of progradation fronts.

We have observed several **indented fronts** in Gabon, notably in front of Conniquet island, south of Owendo harbour and on the Akanda head land (Lebigre, 1990). Erosion, by violently attacking the mud flats which support the mangroves, deeply indents the old front. The result is a winding outline, made of alternate indentations and residual pseudo-headlands. The front itself is characterised by a small embankment shaped in the indurated mud or in the root peat. More recently, we have been able to observe the existence of other examples of this type of front in Mayotte, south of the Boueni bay (Lebigre, 1997). The old *Sonneratia alba* which once formed a pioneer front are now dying and their roots are loose (see Plate 4). In the indentations, these mangroves have completely disappeared and the *Rhizophora mucronata*, which are normally next in line when you move inwards, are exposed directly to erosion.

Fronts presenting residual thickets or individuals are frequent, notably in Gabon (Plate 5). In Madagascar, the mouth of the Sabora creek, in the meridional part of the Tsiribihina delta offers a revealing

Plate 3. Old isolated bf *Avicennia marina* on the strand in Sarodrano (in the South-west part of Madagascar).

Plate 4. Mouth of the Koulale River (Mayotte Island) – peat bank made of compacted clay and rootlets on a retreating front.

Plate 5. Rio Mouni (Gabon) retreating front with residual *Rhizophora racemosa.* Thicket and root peat showing through.

Plate 6. Pointe Idolo (Gabon): straight line front composed of tall *Rhizophora* spp.

Plate 7. Marajo island (brasilian Amazonia): seaward straight line front showing an enbankment slope resulting from erosion and shaped within root peats, and tall *Rhizophora*, dying after the mud flat was invaded by sand.

example of the possibility of reversal of phenomena. On the aerial photographs which were taken in 1954, a diffuse scattering front can be distiguinshed, but on those of 1982, only about 20 big mangroves remain (*Sonneratia alba*). What we have is, therefore, an aborted progradation front, which has transformed into a retreating front. The Passi Keli bay, south of Mayotte, offers a very good example of a mangal constituted only of senescent mangroves, either isolated or grouped in residual copses.

Straight line fronts seem to be due to regular and prolonged erosive phenomena rather than to brutal cataclysms. We have observed some of these phenomena in Brasilian Amazonia near SourÈ (Marajo island), in Gabon, in Mayotte and in Madagascar in the Tsiribihina delta. The most frequent case is the translation from the sea towards the mangal of a chenier at the surface of the mud flat (Figure 4). There are three distinct stages: the dismantling of a coastal bar protecting a sector of mangal, progression of the sand followed by the asphixiation of certain species of mangroves by the sand, the creation of an embankment slope and the uprooting of dead mangroves by the sea (Plate 6).

Some mangroves succeed in surviving the asphixia and the sea's action, which answers for peculiar outlines, intermediate to the retreating fronts presenting residual copses or individuals (Plate 7).

Tanne progress at the expense of the mangal

When a *tanne* develops at the expense of the mangal, the contact between the two is generally marked by the presence of a fringe – of variable width – of dead or dying mangroves, or simply of stumps when timber is exploited by man. This landscape testifies to a retreat of the mangal. It is, however, difficult to know without the help of aerial photographs if such an event has a limited development in time and, if such is the case, to date it precisely: unlike what happens with seaward margins, the sea doesn't knock down the dead mangroves and they can remain standing for a long time unless there is human intervention. Madagascar offers many examples of such margins, but they can also be found in Brasil (Plate 8). In Senegal, Marius (1979) described the retreat of the mangal due to the progress of the herbaceous tanne in *Sesuvium*, due to

Table 1. Summary of the main types of margins

	Retreating fronts	Progradation fronts
Characteristics	– senescent or dead mangroves – root peat outcrops – root loosening – erosion slopes – fallen mangroves	– young mangroves – dense canopy
Seaward fronts types	– fronts with indentations – fronts with residual copses – straight line fronts	– diffuse scattering fronts – front with mangrove copses – crescent fronts
Landward fronts types	– margins with dead mangroves – senescent mangroves margins	– re-colonization fronts

Plate 8. Landward retreating fronts with dead mangroves showing the recent progress of a *tanne* over the mangal in Brasilian Amazonia.

the droughts of 1973 and 1974. Dead mangroves rise above this succulent lawn.

Conclusion

Mangals present a set of very clear examples of margins corresponding to phases of progradation or on the contrary of retreat of the mangal without any human intervention (Table 1). These phenomena have nevertheless become difficult to study in countries where this type of vegetation is exposed to human activities, as in the Philippines for instance. From this point of view, we assume that the typology is far from being complete and can be considerably improved.

The difficulty of making a diagnosis from medium scale aerial photographs (1:40 000) or from spatial images with high resolution (SPOT, LANDSAT TM) is to be noted. In order to reach a correct interpretation, study in the field is compulsory, even though most of the time, this means great transport difficulties.

From a general point of view, the multiplication of studies regarding mangal margins should allow one to measure the real impact of the slow rise of the sea level on various sedimentary and isostatic configurations (Paskoff, 1993).

In addition, it is important to note that the phenomena described above have a strong influence on the stands made on tropical littorals. For example, we can note that many of the failures in restoration operations on mangals are linked to the failure to take into account phenomena of retreat (Saenger et al., 1983; Hamilton & Snedaker, 1984). A good analysis of margins allows one to make a vital diagnosis on the natural spatial dynamics of the parts of the mangal where aquatic developments are to be made. But even this has its limits. Although it is a rare phenomenon, on seaward fronts the sedimentary trend can reverse, while on tanne borders exceptional years in terms of rainfall can also provoke a reversal of trend.

References

Baltzer, F., 1975. Mapping of mangroves in the study of sedimentological processes and the concept of swamp structure. Proc. Int. Symp. Biology and Management of Mangroves, Honolulu: 499–512.

Battistini, R. & J. M. Lebigre, 1994. Milieux-enregistreurs et enregistreurs sur les littoraux tropicaux. In Enregistreurs et Indicateurs de l'Évolution de l'Environnement en Zone Tropicale, Talence, Presses Universitaires de Bordeaux, Coll. Espaces trop. 13: 255–274.

Bertrand, F., 1993 Contribution à l'étude de l'environnement et de la dynamique des mangroves de Guinée. Données de terrain et apport de la télédétection. Paris, ORSTOM, coll. Etudes et thèses: 201 pp. (thèse de doctorat, Bordeaux 3, march 1991).

Hamilton, S. & S. C. Snedaker, 1984. Handbook for mangrove area management. Honolulu et Paris, EPIEWC, IUCN & UNESCO: 123 pp.

Lebigre, J.M., 1990. Les marais maritimes du Gabon et de Madagascar, contribution géographique à l'étude d'un milieu naturel tropical. Université de Bordeaux 3, thèse de doctorat d'Etat: 706 pp.

Lebigre, J.-M., 1994. Les marais maritimes de la baie de São Marcos (Maranhão – Brésil). Travaux du Laboratoire de Géographie Physique Appliquée – Bordeaux 12: 21–35.

Lebigre, J.-M., 1997. Les marais à mangroves du Sud-Ouest de Madagascar – Des palétuviers et des hommes au pays des épines. In Milieux et sociétés dans le Sud-Ouest de Madagascar. Talence, CRET, coll. 'Iles et Archipels' 23: 135–242.

Lebigre, J.-M., 1997. Problèmes d'érosion dans les marais à mangrove de Mayotte (archipel des Comores). Travaux du Laboratoire de Géographie Physique Appliquée 15: 45–58.

Lebigre, J. M. & F. Rebelo Mochel, 1995. Le littoral à mangrove d'Alcântara (Maranhão, Brésil) – Evolution et conséquences relatives à sa gestion. Travaux du Laboratoire de Géographie Physique Appliquée – Bordeaux 13: 23–27.

Lescure, J. P., 1977. La mangrove guyanaise: architecture des jeunes stades et vie avienne. Cahiers de l'ORSTOM, Sér. Biologie, XII, 4: 361–376.

Lescure, J. P., 1980. Aperçu architectural de la mangrove guadeloupéenne. Acta Oecol. Oecol. Generalis 1: 249–265.

Marius, C., 1979. Effets de la sécheresse sur l'évolution phytogéographique et pédologique de la mangrove de Basse-Casamance. Bull. de l'IFAN, Série A 4: 669–691.

Paskoff, R., 1993. Côtes en danger. Paris, Masson: 250 pp.

Saenger, P., E. J. Hegerl & J. D. S. Davies (eds), 1983. Global status of mangrove ecosystems. IUCN (Union Internationale pour la Conservation de la Nature): Commission on Ecology Paper no. 3/The Environmentalist 3: 88 pp.

Hydrobiologia **413:** 115–126, 1999.
R.S. Dodd (ed.), Diversity and Function in Mangrove Ecosystems.

Mangrove dynamics in the Rivières du Sud area, West Africa: an ecogeographic approach

Frédéric Bertrand
PRODIG, Unité Mixte de Recherche 183 C.N.R.S.- Université de Paris I 191 rue Saint-Jacques, 75005 Paris, France

Key words: mangrove ecology, vegetational succession, environmental setting, hydrodynamics

Abstract

Because of the diversity of its mangrove environments, the extensive estuarine coast of West Africa, between Saloum (Senegal) and Sherbro (Sierra Leone) estuaries, provides the biogeographer with a highly favourable framework for the analysis of changes in mangrove communities. These changes can be studied at various organization levels of the mangrove ecosystem, emphasizing the multi-dimensional nature of successional processes. Although being denied elsewhere, the efficiency of biotic processes is demonstrated by numerous case studies relating changes in mangrove communities to simultaneaous edaphic modifications. More recently, disturbances in zonation patterns related to reduction of water discharges draws attention to the fact that varying longitudinal gradients, involving interaction between salt- and freshwater, may give rise to specific successional schemes. Last but not least, how mangroves develop to an equilibrium between deposition in swamp and erosion in the tidal creeks can be appreciated by recognition of a third volumetric dimension of successional processes. Since any successional pattern involves these three dimensional components, an attempt is made to recognize basic 'multi-dimensional' sets of process variables, related to the characteristic depositional settings of the area. As a general rule, it can be stated that successional patterns are all the more dependant on varying environmental gradients since they occur in areas receiving large supplies of allochtonous sediments, either from landward (deltaic estuary) or seaward (tidal bay) origins.

Introduction

The ability of mangrove communities to change by their own is a topical, although 'classical' question. It has never stopped being discussed among mangalogists insofar as new studies were providing various and rather contradictory opinions. However, growing recent concern about reduction of mangrove areas in relation to sea-level rise and human pressure is very likely to confer a renewal of interest in it.

In order to predict long-range directional trends imposed by such changes within an area, it is necessery to develop an understanding of processes involved in successional schemes. As Thom (1984) pointed out about any attempt to comprehensively model a mangrove ecosystem to explain community change, 'the problem is multidimensional in space as well in time, besides being multidisciplinary in scope'.

It is the purpose of this review paper to outline a geographic approach of changes in vegetational patterns emphasizing primarily the three dimensions of physiographic and functional structuration of mangrove ecosystems in which various successional processes are involved. Further attempts will be made to classify a number of major mangrove environment types with each type reflecting a particular combination of successional processes controls. Stability of these basic sets of process variables will be discussed with regard to high degree of climatic variability and conversion which characterize the Rivières du Sud area.

The coastline of this physical as well as human coherent region, 1.135 km long, extends across 7° of latitude from 14° N at the attached end of Sangoumar spit to 7° 30′ at Turner peninsula, from Senegal, Gambia, Guinea-Bissau, Guinea to Sierra Leone. Mangrove swamps cover approximately 9000 km^2 of the coastal zone. The unique character and cohesion of the area stems from tropical-summer rainfall conditions occuring on a depositional setting defined by a drowned bedrock valley system, which has being partly infilled since the Nouackchottian transgressive peak (i.e. 5500–5000 BP). So far, there is no unique description of communities distribution in mangroves. The apparent patterns vary from site to site so that mangroves are more readily interpretable as a mosaic of vegetation types and structures involving varying geophysical features within the area.

Small-scale varying factors affecting mangrove patterns

On a regional scale, differences in spatial pattern development may be interpreted as a function of a broad latitudinal gradient leading from very humid conditions in Sierra Leone ($P/Etp > 0{,}75$) to semi-arid conditions ($0.20 < P/Etp < 0.50$) in Senegal, since it induces subsequent gradual limitations to mangrove landward expansion within the intertidal zone. Nevertheless, as Guilcher (1989) recently pointed out, this gradient doesn't account for the local and sometimes disconcerting diversity of vegetational patterns. As a matter of fact, latitudinal variations of rainwater deficit is in keeping with the general pattern of hydrological cycle within the intertidal basin. Thus, hydraulic balance related to river runoff and tidal exchange appears to be of great significance within the area, although very few data are available. The important point to consider is that tidal range increases from microtidal margins of the area (< 2 m spring range) to its macrotidal middle part (spring range up to 7 m in Rio Geba and Rio Grande de Buba) involving simultaneous decrease of wave energy to the shoreline (see Figure 1).

Following the mangrove environment classification of Woodroffe (1992) (cf. also Thom, 1984) and bioclimatic classification of Blasco (1984), three geophysical backgrounds may be roughly distinguished within the area on the basis of oceanic and bioclimatic features: the northern semi-arid and wave-dominated coast, the central subhumid and tide-dominated coast, the southern very humid and composite river and wave dominated coast.

This three-fold regional division of the Rivières du Sud accounts for varying diversity of zonation patterns with highest degree of diversity in the central region – related to a great array of geomorphically-defined habitats – and lower degree of diversity in the flanked regions, especially along the 'autochtonous' northern coast where sediment influx is presently limited (Bertrand, in press).

The three dimensions of successional processes in mangrove ecosystems

The transversal dimension of vegetational effects

Since the frequent clear-cut zonation of mangroves within the tidal flats provides the investigator with an accurate visual framework for observing changes in communities, the transversal dimension of mangrove dynamics has been largely documented. To some extent, satellite imagery and aerial photographs have provided some more stimulus for studying mangrove sequences by complementing the transect method by a plan view, as in Senegal (Diop, 1986) and Guinea (Bertrand, 1993). Yet, the relationship between zonation and succession within the area has not been clearly discussed even though it is a very common topic in mangrove ecology (Snedacker, 1982). A reason is that West-African mangalogists might tend to get confused between two basic concepts.

The toposequence concept is conventionally used in many works (Baltzer & Lafond, 1971; Blasco, 1983; Bertrand, 1993) to relate typical zoned patterns to the microtopography of tidal flats. Following Watson's early ideas in this field (1928), mangrove communities are considered to occupy separate topographic sections of the tidal range insofar as salinity is correlated to tidal and hydroperiod characteristics. With such a concept, zonation is related to degrees of surface inondation which are physiologically significant. Therefore, vegetational belts may be correleted to inondation classes reflecting the transitional status rather than the successional status of mangrove communities.

The serial vegetation concept provides a different, albeit often overlapping approach of mangrove communities relating observed zonation patterns to successionnal processus. Mangrove ecologists have made extensive use of this concept within the Rivières du Sud area since french botanist Schnell's idea of mangal associations representing a successional stage in the development of some climax *Rhizophora*-dominated communities after destruction of the previous ecosystem (Schnell, 1951). Pedologists (Chezeau & Jacques-Félix, 1960; Jordan, 1964) have played an important role in generalizing ecological analyses of mangrove succession by emphasizing vegetational influence on

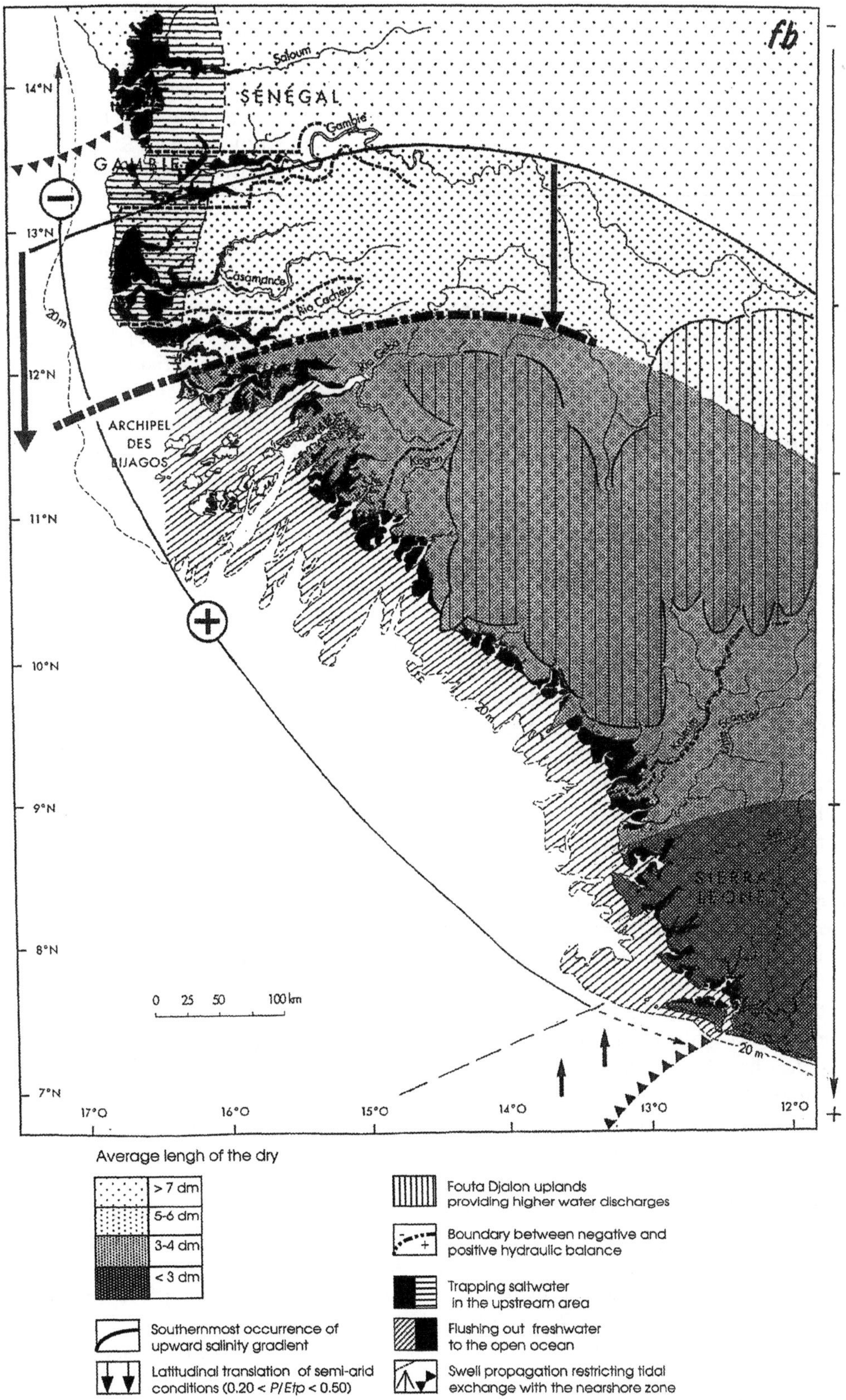

Figure 1. Regional geophysical components of mangrove environments within the Rivières du Sud area

soils and pedogenesis. The main emphasis of their works is that changes in mangrove communities cannot be understood without taking into consideration previous modifications in both morphological (ripening) and chemical (acidification, salinization) soil characteristics related to mangrove species properties. Because of the varying ability of species to induce organic (autochtonous) sedimention and to trap sulphur through their root system, pioneer species established in primary successions (on unvegetated bare tidal flats) may modify the habitat in a more or less favourable way for successors. In this approach, zonation in space represents a serial succession of vegetation in which development seems to be under the control of the 'sequence-head' species. It is noteworthy that, in numerous areas where disturbances in the vegetation pattern are introduced by mangrove utilization (such as in Guinea), generalized zonal succession schemes may be used as autogenic successional models to predict direction of changes in communities (Bertrand, 1993).

A related advance of great potential in the whole understanding of successional processes has come from the concept of 'chronoséquence' as it was formalized by Viellefon in the early 1970s in Senegal. According to this pedologist, it is not one of these external (tidal range) and internal (pedogenesis) factors but their effect in concert that establishes the characteristic mangrove-tanne pattern within this tropical area. Zonation represents a series of pedological and vegetational units which are genetically related to one another and operate together as a complex unity. This type of sequence whose most famous one is the Balingore sequence shows contrasting patterns of development in relation with successive predominant facies of mangrove within the studied area. Therefore, time ('chrono-') is the key factor which links the vegetional effects to patterns of spatial organization.

Viellefon's ideas have been widely diffused, especially throughout the Western mangal floristic region where the very low number of true mangrove species, i.e. no more than eight, gives more facilities for relating changes in mangrove communities to simultaneaous edaphic modifications. However, this synthetic approach of mangrove dynamics was reduced to being relevant to only parts of the Rivières du Sud area which are characterized by 1. a high content of silt or clay material in the substrate and 2. a high level of geomorphic stability (Marius, 1985; Anthony, 1990; Bertrand, 1993). The first condition is rarely found near the mouths of rivers and near the landward margins of mangals where sand delivery from either nearshore or terrestrial sources may provide sandy substrates in which strong groundwater flux engenders poor zonation and reduces vegetal dynamics. The second condition is almost absent along the central part of Northern Rivers (e.g. between Rio Geba in Guinea Bissau and the Conakry peninsula in Guinea) where a combination of high tidal range (> 4 m) and high seasonal river discharge lead to continuous geomorphic change over time. The result is high diversity of ecophysical habitats and spatial patterns of mangrove development involving a wider range of pedogenic dynamics (Bertrand, 1995) and more frequent disturbances in serial succession of vegetation (Bertrand, 1993). In many cases, it clearly appears that neither organic (autochtonous) nor inorganic (allochtonous) accumulation around the mangrove root system is capable of over-riding abiotic land-building processes.

This limitation for generalizing further ecological analyses of mangrove zonation into chronoserial succession models demonstrates the necessity for a more comprehensive understanding of zonal successional schemes to consider the longitudinal dimension of the problem.

The longitudinal dimension of river to saltwater transition effects

A further consideration in understanding the factors controlling mangrove dynamics is to take into account the mixing regime of the fresh- and saltwater masses in estuaries.

Up to the 1960s, hydrosedimentological processes and estuarine hydrology had been considered as steady factors because river regime appeared regular even when the northernmost part of Northern Rivers (such as the Casamance river) experienced annual fluctuation in discharge related to variability in precipitation. But since the annual rainfall started to decrease in the early 1970s, inducing a reduction of freshwater inputs throughout the whole area, these factors were found to be variable. Meanwhile, some spectacular changes in vegetational patterns and species composition started to be observed, especially in Senegal (Marius, 1985). It follows that mangrove dynamics might be strongly influenced by allogenic processes related to varying longitudinal gradients.

Interaction between salt- and freshwater gives rise to various biological process and patterns within the area. Depending on the relative amount of river discharge and tidal input, varying salinities may be more pronounced in the middle part of the estuary, as in the Gambia River (Marius, 1985), in the upper part as in the Casamance River (Marius, 1985) or in the lower part as for the Konkouré River (Berthois, 1967). Since mangroves have to adjust to such varying salinities for their whole life, longitudinal salinity gradient often produces a longitudinal zonation of predominant species. In Guinea for instance, an opposition can be easily recognized between an estuarine fluvial zone dominated by species which tolerate a low range of salinities (*Rhizophora racemosa* and *Rhizophora harrisonii*) and estuarine marine zone dominated by species which tolerate a higher range of salinities (*Rhizophora mangle*, *Laguncularia racemosa* and *Avicennia germinans*). In areas where natural patterns have been broadly disturbed, as the Soumbouya River south of Conakry, herbaceous secondary communities may provide an even more distinctive pattern of longitudinal zonation with the appearance of a middle zone characterized by the dominance of *Philoxerus* (Baltzer & Lafond, 1971; Bertrand, 1993). Although herbaceous plants are not inhabitants of strict mangrove communities, their longitudinal zonation emphasizes the influence of longitudinal salinity gradient on successional processes.

This influence turned out to be essential when river runoff started to decrease throughout the area in relation with the climatic drought. As Pages et al. (1987) pointed out in Casamance this led to a raising of salinity profile upstream. Depending on the latitudinal location of the estuary and its drainage basin geometry, the seaward gradient lessened or reversed to a landward gradient (as shown in Figure 2). In both cases, the increasing of average salinity of tidal waters resulted in the replacement of *Rhizophora* species by *Avicennia*, especially within the higher intertidal tidalflats. But in several upper reaches of estuaries, as in Middle and Upper Casamance and in many drowned tributary valleys where former varying salinites were already closed to the level tolerated by the most euryhaline species (40 ‰), levels of salinities – in both variation and average – became so high throughout the intertidal zone as to be physiologically insignificant in terms of zonation effects. As a consequence, successional patterns which were previously based on mangrove properties are broadly disrupted and even replaced by patterns much more dependant on unstable hydroclimatic conditions.

The implications of characteristic mixing regimes on mangrove successional patterns are poorly docu-

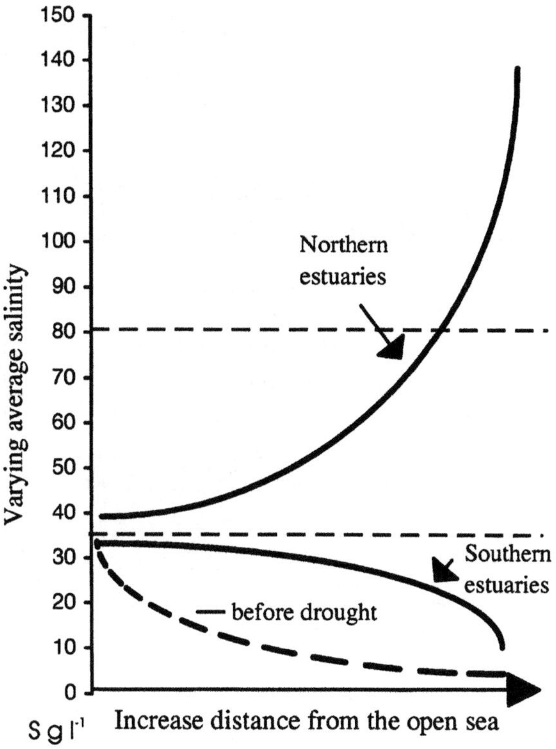

Figure 2. Comparative modifications of longitudinal salinity gradients due to decreasing freshwater inputs.

mented. Yet there now exist a number of case-studies which highlight the importance of understanding hydrosedimentological processes in the interpretation of temporal variation in mangrove communities (Anthony, 1990; Bertrand, 1993). Because of varying river discharge and tidal range, the three basic mixing regimes, i.e. stratified, partially mixed and fully mixed types, are likely to occur. The Saloum River is a classical salt wedge, although inverse estuary, whereas the Scarcies rivers and Rokel river are good examples of mixed estuary (Anthony, 1990). Since each type produces different kinds of residual currents, surrounding mangrove may be subjected to varying hydrologic energies influencing their zonation and succession pattern. No systematic study of vegetational patterns have been carried out in relation to mixing regime. However, using both infrared aerial photographs and available hydrodynamic data, it is possible to correlate mangrove habitats to various modes of mixing.

In stratified estuaries such as the Konkouré (Berthois, 1967), where the outflow of fresh- to brackish water on the surface is compensated for by an inflow of seawater near the bottom, two major habitats may often be distinguished in association with depositional

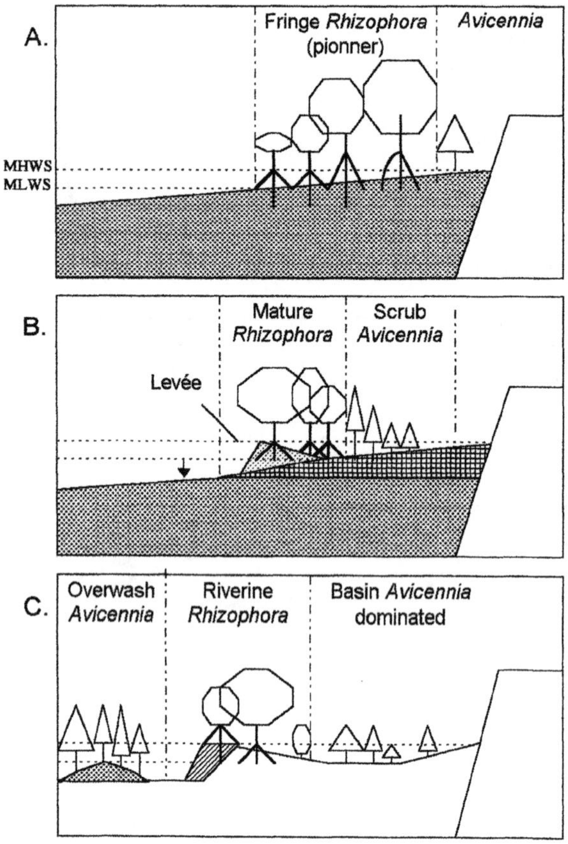

Figure 3. Schematic diagram to show the pattern of mangrove succession within a stratified estuary, based on the Konkouré in Guinea

landforms (Bertrand, 1993). Riverine *Rhizophora*-dominated forests line the overbank or levee deposits formed when bankfull discharge leads to the deposition of fine sediment beyond the channel itself (see Figure 3B, C). Overwashed *Avicennia*-monospecific forests colonize channel bar deposits on the convex side of meander bends (see Figure 3C). At a higher level of observation, the distributional pattern of these two distincts habitats suggests a determining influence of river mouth processes which may well be controlled by friction, as is often in such shallow waters (Wolinsky, 1990).

In contrast, where tidal range and current velocities are large enough to breakdown the vertical salinity gradient, tidal circulation may override density circulation, giving rise to different sedimentological and vegetational patterns. Anthony (1990) emphasized the circulatory tidal patterns which may occur in funnel-shaped well-mixed estuaries such as Sherbro Bay in Sierra Leone under the influence of geostrophic forces: the flood tide tends to be deflected to the left bank with ebb occupying the opposite right bank. Elements of such a complex circulation include a marked oblique gradient in the surface salinity structure and evasive dispersal routes leading to fine sedimentation on the riverine-dominated right bank and coarser material sedimentation on the opposite marine-dominated bank. The implications of these lateral variations on mangroves communities can be analysed in the Sangaréya Bay in Guinea where a very similar surface salinity structure has been described (Nikolayenko & Diane, 1988). On the north flank of the bay, Khonibompé island exhibits a typically progradational shoreline gradually building out in the estuary and fringed by *Avicennia* monospecific forests. Steep elevation gradient does not lead to an apparent zoning of species because muddy sediments accrete rapidly to an elevation at which they get are inundated by tides, involving a second reduced mangrove zone finally transitional to bare flats beyond the influences of tides. In such a depositional system, zonation is a series of regular community structures parallel to the coastline, with mangroves following the migration of their habitat seaward. The opposite flank of the bay exhibits a somewhat different progradational bayhead chenier plain, involving alternations between muddy progradation and episodic beach ridge formation at approximately mean high-tide level. Under conditions of efficient beach drifting, barrier deposits prevent mangrove communities from extending seaward, often involving a switch from open-ocean to overwashed back barrier habitat. But since residual current direction leads to a net flood transport on this flank of the bay, a recurved sandy spit is likely to form at the distal – unattached – end of the barrier, allowing a landward *Avicennia* fringe forest to develop.

Such relationships between mixing regimes and variations, both lateral and longitudinal, in successional schemes already provide a more comprehensive understanding of mangrove dynamics as a whole. However, the tidal wave propagation in the estuary cannot be considered independantly of the densely vegetated surrounding swamps where part of the water flows. In addition, to a longitudinal approach, it is necessary to have a cross-sectional view of succession.

The volumetric dimension of tidal asymmetry effects

A key for developing a global, multidimensional perspective of mangroves dynamics should be to take into account the modification of tidal hydrodynamics from

A.

B.

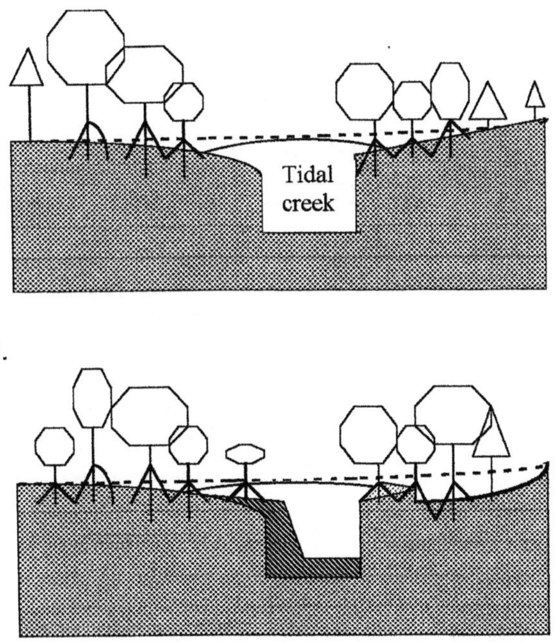

Figure 4. Schematic cross-sectional diagram of a mangrove swamp near Singuéya Island, in Guinea, illustrating the response of mangrove communities to the retardation of tidal flow (adapted from Wolinsky, 1995).

a flood-dominant to an ebb-dominant system, due to the retardation of the tidal wave by the large frictional effects caused by the vegetation (Wolinsky et al., 1980). The determining role of mangrove swamps in the maintenance of tidal basin geometry has already been demonstrated in other mangrove areas: larger ebb currents prevent tidal creek from siltation induced by a net landward transport of particules within the nearshore zone (Wolinsky, 1995). We found in Guinea that ebb-flood asymmetry had implications on successional processes also (Bertrand, 1993). While tidal creeks are clearly adjusted to discharging the mangrove tidal prism, swamp surfaces are adjusted to the retardation of tidal flow. The retention of the flood-tide waters in the swamp leads to a progressive infilling, resulting in cross-sectional migration of mangrove habitats (see Figure 4). *Rhizophora*, in particular, may follow channel margin infill to constitute a wide fringe forest inundated and flushed by all high tides. Where bankfull conditions remain frequent, such as on concave sides of meander bends, most of the sediment is accumulated on the backswamp surface, often involving a switch from riverine *Rhizophora* forest to mixed *Avicennia* and *Rhizophora* basin forest. A highest sedimentation towards the innermost parts of the swamp may be marked by dwarf forms of both species.

These observations suggest that ebb-flood tidal asymmetry is a key factor governing the successional pattern of mangrove communities. However, it should be kept in mind that an extremely dense mangrove vegetation tends to reduce the ebb flow dominance in the creek (Mazda et al., 1995). Therefore, it should be stated that this successional pattern will develop until an equilibrium point between deposition within the swamp and erosion in the tidal creeks, around the level at which seaward dispersion of materials is declining.

Enlarging the finding of Mazda et al. (1995), we suggest that such adjustments between process and form involve a process-response system which characteristic is negative feedback, i.e. a condition whereby the mangrove ecosystem is capable of adjusting in a way that minimizes the effect of species properties on the sediment budget dynamics.

Environmental settings for mangrove dynamics

Succession may be studied further at each of these three organization levels involving different variables. But since no one of these variables establishes the successional pattern by itself, an attempt has to be made to recognize basic 'multi-dimensional' sets of process variables, related to the characteristic depositional settings of the area (see Table 1).

Chenier plain environments: prevalence of allogenic processes

Anthony (1990) has published a detailed analysis of a chenier plain situated astride the frontier between Sierra Leone and Guinea, and whose depositional units are very characteristic of the bight-coast plain stretching from Bullom Peninsula in Sierra Leone to Kaloum Peninsula in Guinea. Although this author focuses on modes of chenier formation related to morphodynamic changes, he raises some of the questions concerning succession that we further discussed within similar environmental settings in Guinea (Bertrand, 1993). As a result of episodic building of chenier ridges in an otherwise rapid progradation dominated by muddy sedimentation, intertidal muds are covered by extensive *Avicennia germinans* fringe forests, whereas *Rhizophora*-dominated communities only form narrow strips along the inner creeks. On this

Table 1. Ecologic responses to varying combinations of environmental process variables. Some attempt is made to quantify these responses and the relative importance of processes (H = High, M = Moderate, L = Low; ++ = very active, + active, (+) = optional). Note that climatic conditions are not directly mentioned

Depositional pattern	Dominant physical process			Longitudinal mixing regime	Ecological response		Example
	River	Tide	Wave		Zonation diversity	Community stability	
Barrier estuary		(+)	+	Stratified (locally inverse)	L	H	Saloum Lower Casamance
Bight-coast chenier plain		+	+	Variable	L	L	Kaback, Kakossa Northern Sierra
Deltaic estuary	++	+		Partially mixed	H	M	Konkouré
Estuarine delta	+	++		Well-mixed	M	L	Mellacoree, Scarcies
Ria estuary	(+)	++		Partially to well-mixed	M	M	Gambia Rio Grande de Buba
Tidal bay		+		Well-mixed	H	L	Sangaréya Bay Sherbro estuary

geomorphologically active shoreline where autochtonous sediments (e.g. those which are deposited *in situ* including mangrove-peats) are limited, changes in habitats and plants occupying these habitats are clearly found to occur subsequent to sedimentation on seaward flats.

Where seasonal rainfall remains high for a sufficiently long period (at least one or two decades), assuring an active supply of muds from upstream, the natural tidal flooding is bound to cease rapidly on the inner margins of the fringe forest. As a consequence, mangrove dieback will result in the formation of bare supratidal flats called 'tannes' in the case of no water supply during the dry season, or supratidal freshwater swamps where abundant ground-water seepage may balance summer evaporation. Both cases show an accelaration of successional processes under rapid progradation conditions.

Conversely, higher-energy wave conditions associated with either lower rainfall (Bertrand, 1993) or strengthened southwesterly winds often leads to a successional truncation and even elimination in place, especially where its landward edge has been cleared for rice farming or salt production as in Koba (Bertrand, 1993). As Semeniuk (1980) already noticed in a comparable eroding shore in Western Australia, interior stands of mature *Avicennia* may emerge by dint of erosion as the seawardmost vegetation. Meanwhile, erosional sand-winnowing from the mudflats and mangrove fringe entrapment of washover deposits (including also shore-supplied sand) result in the formation of closely spaced cheniers up to the extreme high-water level. Suprisingly, mangrove burying under deposition, including its above-ground root extensions, does not always induce changes in landward habitats. The reason could be linked to the well-known ability of *Avicennia* to resist on medium to coarse-textured substrates. But the observation of seedlings lodging, geminating and growing on the backslope of many washover deltas also suggests an ability such communities to override changes in progradation mode (i.e. from the rapid to the 'cut and recover' mode) and local character of sedimentation (i.e. muddy to sandy sedimentation). Whatever prograding pattern may prevail, changes in vegetation patterns and species composition within chenier plain settings appear to be much more related to abiotic land-building processes than to simultaneous pedogenetic modifications.

Barrier estuary environments: prevalence of autogenic processes

It is somehow exactly the opposite situation that can be observed within bay barrier environments in which tidal lagoons occupy drowned river valleys impounded by coastal sand barriers. Due to its closed association with both micro- to medium tidal range (< 2, 5 m) and high wave energy at the shoreline, this terrigenous setting typically occurs in the northermost and southernmost parts of the Rivières du Sud where they are very well documented (Marius, 1985; Anthony, 1990).

Although water bodies entraped behind coastal barriers show varying degrees of infilling (depending on coastal evolution and actual rate of fluvial input), they are all characterized by narrow and shallow tidal inlets connecting larger estuaries to the ocean. Many studies showed that in such settings, mangrove distribution is dependant on the efficiency of bidirectionnal currents within the river mouths leading partly to fully mixed estuaries under low flow conditions (Thom, 1984). As a matter of fact, the resulting residual circulation is clearly in favour of a net landward transport of particles in tidal estuaries and a subsequent mangrove fringe progradation. Yet, most of barrier estuaries, as Casamance or Saloum Rivers, exhibit partial infilling and mangrove communities which ought to be relevant to riverine forest type rather than to fringe forest type, according to the classification of Lugo & Snedaker (1974). It could be argued that these barrier estuaries did not reach the infilled stage since mangrove progradation started only 3900 or 3500 years B.P. (Faure et al., 1974). But considering that climate has tended to become globally less humid meanwhile, we would rather consider this partial infilling as a longterm trend reflecting progressively declining sediment supply within embayments in relation to decreasing water discharge.

Under such steady-state conditions in which there is a global balance of erosion and deposition within the sheltered barrier esturies, changes in vegetation patterns and species composition on stabilized tidal mudflats give evidence of the efficiency of biotic processes. Abundance of marine sulphates, due to exeeding seawater input- and dominance of *Rhizophora* species, whose well developped root system constitutes a highly effecient trap for sulphur favouring the ripening of morphological and geochemical properties of mangrove soils in relation to varying salinities (Marius, 1985). Development of consistency in turn affects primary production and plant-species distribution by reducing the lateral permability which pemitted tidal influences to occur far beyond drainage channels and, therefore, to damp out the effects of potential evapotranspiration (*Etp*). Thus mangrove communities are very likely to represent successional stages in the development of some climax. This has meaning only relative to position along cross-sectional gradients that may change over time in relation to rainfall variability, especially in the northernmost part of the Rivières du Sud where semi-arid conditions ($0.25 < P/Etp < 0.50$) prevail.

It is noteworthy that relative steady-state equilibrium within barrier estuary settings does not prevent mangrove communities from cyclic short-term instability in relation to varying degrees of aridity. In addition, we suggest that instability may be enhanced by overwashing seawater accumulating in the barrier estuary since several bay barriers such as Sangomar spit enclosing Saloum estuary in Senegal and St. Ann peninsula enclosing the wide Sherbo estuary in Sierra Leone appeared to experience overwash breaching on different time-scales (Anthony, 1990; Diaw et al., 1990). Up to now, attention has been focused on breaching mechanisms and subsequent inlet formation, as for example on Sangoumar barrier near the Lagoba breach. But, no doubt such changes in the barrier geometry also affect environmental gradients within the lagoon, especially by upsetting its hydrochemical balance. While this may be so, involving only adjustments of successional schemes to decreasing or increasing gradients, persisting *Rhizophora*-dominated communities within the barrier estuaries might be the result of a dynamic equilibrum including varying physiological responses to temporary hydrochemical change rather than from a steady-state one involving no change over the period.

Intermediate environmental settings

Beside these two basic and somehow extreme environmental settings, the Rivières du Sud area show settings in which mangrove dynamics is clearly related to composite-allogenic and autogenic processes. Under such indefinite successional schemes, patterns of mangrove species distribution and structural types may be disrupted by any single modification of environmental conditions. Between all factors which may affect direction and magnitude of mangrove dynamics within these 'intermediate' environmental settings, there is a general belief among West African mangalogists that varying water discharge related to unstable climatic conditions and multiple-purpose use of mangrove ecosystems are the most efficient external factors.

Successional disturbing effects of recent great drought

Most of the studies emphasized the effects of consecutive and pronounced water deficits on species zonation since estuaries, especially in the Sahelian

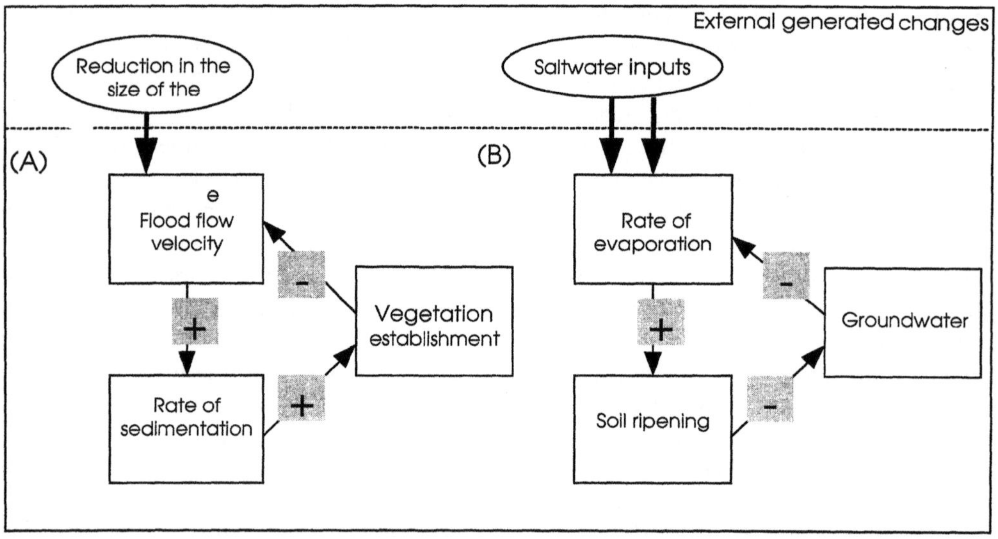

Figure 5. Examples of negative (A) and positive (B) process-response systems involved in mangrove successional schemes. Negative and positive signs indicate the nature of relationship between variables.

northern part of the area, have been experiencing drastic reduction of freshwater inputs since the early 1970s. However, little attention has been paid to the environmental setting in which most spectacular disturbances occured, such as in the middle and upper part of Casamance River (Marius, 1985) or in the Bintang Bolon River in Gambia (Blasco, 1983). Now we draw attention to the fact that these situations were observed in very similarria estuary environments in which low amounts of river discharge may lead, in case of consecutive rainfall deficits, to isolating the upper reaches of the estuary from less saline coastal waters. Under subsequent hypersaline and evaporatic lagoonal conditions with salinities exceeding 80 ‰ (Pages & Debaunay, 1987), disappearance of mangrove stands on the inner tidal flats results in a decreasing proportion of organic matter contained in soil which ultimately favours its compaction (Vieillefon, 1969). Unequal deposition of this kind within the inter-tidal zone leads to a 'concave' deposition pattern, disrupting the discharge of spring tidal waters and ground-water to the drainage channels. At this state of the successional scheme, antagonistic influences of rainfall and evapotranspiration may play an overwhelming role in accelerating, rather than decelerating, the whole process through direct physiological effects on vegetation.

Finally, rapid extension of decaying mangrove and tidal bare flats at the expense of riverine forests is likely to exhibit a positive feedback system where seawater input engenders a system response which produces output due to evaporation, which in turn reinforces the original input and ultimately causes a shift in the mangrove ecosystem to a new equilibrium state. Therefore, ria estuary should be recognized as a major environmental setting within the Rivières du Sud area because of its unique tendancy to amplify magnitude of changes in communities related to climate deterioration by comparison to direct biological implication of drought on species distribution (see Figure 5A).

Successional disturbing effects of mangrove swamps reclamation

Man-made successional disturbances are less documented, probably because traditional conversion of mangrove soils throughout the area has led to more attention given to subsequent pedogenetic modifications than to responses of mangrove communities. In other respects, Bertrand (1995) recently discussed interactions between vegetation and soil components of mangrove ecosystems within the area. This overview came to the conclusion that spatial pedologic heterogeneity is strictly controlled by species distribution at early stages of pedogenesis, but that vegetation communities also exhibit a relative self-governing behaviour with regard to pedogenetic evolution.

The latter feature of the relationship is demonstrated by secondary successions occuring after partialy or entire destruction of previous communities and leading to similar vegetational patterns (Bertrand, 1993). In many areas, such as the Bignona and Guidel

Rivers in Casamance (Marius, 1985), the Rio Mansoa in Guinea Bissau or the Yomboya plain in Guinea (Bertrand, 1993), *Avicennia* propagules appear to disseminate and germinate over abandoned reclaimed soils in which original 'sulfidic' material related to former *Rhizophora*-dominated communities has undergone drastic and very significant changes to mature acid sulphated soils. This means that changes in successional patterns widely override changes in the morphology and geochemistry of mangrove soils, obeying more directly unstable inter-tidal hydrology.

It is only where reclamation procedures involve the enclosure of mangrove areas, and therefore a disruption of the tidal regime, that successional schemes are likely to be disturbed (Bertrand, 1993). However, since the area has a long but rather 'turbulent' history of land reclamation for rice production (Cormier-Salem, in press), significant man-made disturbances of successional patterns are quite difficult to identify in the long term. As a general rule, traditional techniques consisting in temporary (seasonal) and space-limited infringement of the tidal flow do not affect radically the hydrodynamics (Bertrand, 1993). But in Guinea where extended areas, up to several thousands of ha each, have been empoldered in order to increase the rice production, disruption of the tidal regime led to substantial hydrodynamic and morphologic changes. In some cases, e.g. in Monchon area and the upper Kapatchez estuary, the digging of ditches to improve the drainage of mangrove backswamps shifted the balance of forces towards a pronounced tide domination, accentuating the mean sea-level within the whole reclaimed area. As a result, mangrove species, especially *Avicennia*, started to colonize former supratidal flats which were previously covered by back mangal communities.

In other areas, where only the upstream part of the tidal basin was empoldered, the unreclaimed part of the estuary may have markedly decreased in size, involving a spectacular extension of *Rhizophora* riverine forest, as for example in Dofili River near Conakry. In such a case, we can hypothesize that reduction in tidal prism volume within the estuary induced a more symmetrical form of tidal curve (due to the loss of the retarding zones) which, in turn, reduced scour and promoted sedimentation until the frictional effects of the riverine forest could re-establish the ebb-flow dominance. Under these circumstances, the mangrove structure can be said to be in steady-state equilibrium in the long term, since no change of the ratio of the swamp area to the creek area over the period in question means that the structure is capable of adjusting in a way that minimizes the effect of externally generated changes (as shown in Figure 5B).

Conclusion

Considering the high diversity of mangrove environments within the studied area, the successional status of mangrove communities has to be related to varying combinations of environmental process variables. This overview shows that an application of process-response system analysis may lead to some substantial advance in understanding interactions between biotic and abiotic processes, by emphasizing the hierarchical property of the mangrove ecosystem. Thus, it is possible to define the relative importance of different state variables within each environmental setting.

As a general rule, it can be stated that successional patterns are all the more dependant to varying environmental gradients since they occur in areas receiving large supplies of allochtonous sediments, either from landward (deltaic estuary) or seaward (tidal bay) origins. Under these circumstances, there is little chance to find steady-state communities, and we may add that direction and magnitude of changes in communities through time are rather unpredictable.

Conversely, the successional status of mangrove communities is easier to identify within environments characterized by a high level of energy dissipation (barrier estuaries), where changes in communities can be seen in the context of long-range directional trends. In such conditions, the efficiency of biotic factors on successional schemes and the ability of mangrove communities to ensure a long term self-maintenance is directly related to the stability of the background geophysical components (climatic change and sea-level history).

References

Anthony, E., 1990. Environnement, géomorphologie et dynamique sédimentaire des côtes alluviales de la Sierra Leone, Afrique de l'Ouest. Rev. Anal. Spat. 27–28: 93 pp.

Baltzer, F. & L. R. Lafond, 1971. Marais maritimes tropicaux. Rev.Geogr Phys. Géol. Dyn. 2: 173–196.

Barusseau, J. P., E. S. Diop, J. L. Saos & P. Giresse, 1986. Caractères hydro- dynamiques et conséquences géologiques du fonctionnement inverse de certains estuaires tropicaux. 5th Eur. reg. meet. Sediment.: 45–46.

Berthois, L., 1967. La sédimentation en période de crue du fleuve Konkouré. Rev.Géog. Phys. Géol. Dyn.: 253–262.

Bertrand, F., 1993. Contribution à l'étude de l'environnement et de la dynamique des mangroves de Guinée. Données de terrain et apport de la télédétection, Collection Etudes et Thèses, Paris, ORSTOM, Paris: 201 pp.
Bertrand, F. 1995. Les relations sols / végétation dans les mangroves des pays des Rivières du Sud. In Cormier- Salem, M. C. (ed.), Dynamique et Usages de la Mangrove dans les Pays des Rivières du Sud. Paris, ORSTOM, Coll. Colloques et séminaires: 57–64.
Bertrand, F., 1999. Les Rivières du Sud: une unité éclatée, In Cormier-Salem, M. C. (ed.), Sociétés et mangroves des Rivières du Sud, ORSTOM, Paris: 319–378.
Blasco, F., 1983. Mangroves of the Gambia and Senegal. Ecological status – evolution. Toulouse, I.C.I.V.: 82 pp.
Blasco, F., 1984. Climatic factors and the biology of mangrove plants. In Snedacker, S. C. & J. G. Snedacker (eds), The Mangrove Ecosystem: Research Methods. UNESCO: 18–35.
Chezeau, R. & H.Jacques-Félix, 1960. Sols et groupements végétaux de la zone littorale de Guinée dans leur rapport avec la riziculture. Agror. Trop. 18: 325–341.
Cormier-Salem, M. C (ed.), 1999, Sociétés et Mangroves des Rivières du Sud, ORSTOM, Paris: 416 pp.
Diaw, A. T., N. Diop & Y. F. Thomas, 1990. The rupture of the spit of Sangoumar, estuary of the Saalum, Senegal. Washington, Progress in Aeronautics and Astronautics 128: 170–180.
Diop, E. S., 1986. Estuaires holocènes tropicaux. Etude de Géographie physique comparée des 'Riviéres du Sud' du Saloum à la Melacorée. Thèse Doc. Etat ès lettres, Université Louis Pasteur, Strasbourg, 2 tomes: 498 pp.
Faure, M., P. Elouard & S. Diop, 1974. Evolution de la ligne de rivage holocène en Casamance. Bull. A.S.E.Q.U.A., no 42–43: 91–99.
Guilcher, A., 1979. Marshes and estuaries in different latitudes. Interdisc. Sci. R. 4: 158–168.
Guilcher, A., 1989. Are mangroves good tests in coastal geomorphology zonality ? Geogr. Arbeiten 18: 125–148.
Jordan, H. D., 1964. The relation of vegetation and soil to the development of mangrove swamps for rice growing in Sierra Leone. J. Appl. Ecol. 1: 209–212.
Lugo, A. E. & S. C. Snedacker, 1974. The ecology of mangroves. Ann. Rev. Ecol. Syst. 5: 39–64
Marius, C., 1985. Mangroves du Sénégal et de la Gambie. ORSTOM, Paris, Travaux et Documents ORSTOM, no 193: 516 pp.
Marius, C., 1986. Environnement et sols des estuaires de la Guinée Bissau, Rapport de l'excursion du 3e Symposium International sur les sols sulfatés acides. ORSTOM, Paris: 27pp.
Mazda, Y, N. Kanazawa & E. Wolinsky, 1995. Tidal asymmetry in mangrove creeks. In Wong, Y. S. & N. F. Y. Tam (eds), Asia-Pacific Syposium on Mangrove Ecosystems. Hydrobiologia 295: 51–55
Nicolayenko, E. G. & L. Diane, 1988. Structure thermohaline des eaux de la baie de Sangaréah. Conakry, Cerescor 4: 18–40.
Pages, J., J. P. Debenay & J. Y. Le Brusq, 1987. L'environnement estuarien de la Casamance. Rev. Hydrobiol. trop. 20: 203–218.
Schnell, R., 1952. Etude phytosociologique et phytogéographique. Mélanges botaniques Monod. Mém. IFAN 18: 83–217.
Semeniuk, V., 1980. Mangrove zonation along an eroding coastline in King Sound, northwestern Australia. J. Ecol. 68: 789–812.
Snedacker, S. C., 1982. Mangrove species zonation: why ? In Sen, D. N. & K. S. Rajpurohhit (eds), Tasks for Vegetation Science. The Hague, W. Junk: 111–125.
Thom, B. G., 1984. Coastal landforms and geomorphic processes. In Snedaker, S. C. & J. G. Snedaker (eds), The mangrove ecosystem: research methods. UNESCO: 3–17.
Viellefon, J., 1969. La pédogénèse dans les mangroves tropicales. Un exemple de chronoséquence. In Sciences du Sol, Supl. au Bull. Ass. Fr. Et. du Sol: 115–148.
Watson, J. G., 1928. Mangrove forests of the Makay Peninsula. For. Rec. 6: 1–275.
Wolinsky, E., M. Jones & J. S. Bunt, 1980. Hydrodynamics of a tidal creek- mangrove swamp system. Aust. J. mar. Freshwat. Res. 31: 431–450.
Wolinsky, E., Y. Mazda & P. Ridd, 1990. Mangrove hydrodynamics. In Robertson, A. I. & D. M. Alongi (eds), Tropical Mangrove Ecosystems. American Geophysical Union, Washington, DC: 43–62.
Woodroffe, C. D., 1992. Mangrove sediments and geomorphology. In Robertson, A. I. & D. M. Alongi (eds), Tropical Mangrove ecosystems. American Geophysical Union, Washington, DC: 7–41.

Hydrobiologia **413**: 127–133, 1999.
R.S. Dodd (ed.), Diversity and Function in Mangrove Ecosystems.

Mangrove mapping in North-Western Madagascar using SPOT-XS and SIR-C radar data

Vanina Pasqualini[1,2], Jacques Iltis[3], Nadine Dessay[1], Marc Lointier[1], Olivier Guelorget[4] & Laurent Polidori[1]
[1]*IRD, Maison de la Télédétection, rue J.F. Breton, 34093 Montpellier, France*
[2]*Université de Corse, Equipe ≪Ecosystèmes Littoraux≫, B.P. 52, 20250 Corse, France*
Tel: [+33] 4 95 45 00 55; Fax: [+33] 4 95 46 24 41; E-Mail: pasquali@univ-corse.fr
[3]*IRD, Laboratoire d'Etudes Rurales, B.P. 5045, 34032 Montpellier Cedex, France*
[4]*Laboratoire d'Hydrobiologie Marine, UMR 5556 CNRS, Université Montpellier II, CC 093, Pl. E. Bataillon, 34095 Montpellier Cedex, France*

Key words: mangrove, Madagascar, mapping, remote sensing, SPOT-XS, SIR-C radar

Abstract

Within the framework of an environmental impact study of aquaculture activities in the coastal marsh of Mahajamba (North-Western Madagascar), an assessment of the environmental conditions existing prior to the onset of aquaculture activities (production began in October, 1993) is essential. To this end, the area was mapped using SPOT-XS data, which made it possible to define the main ecodynamic tendencies of the mangroves before the development of the rearing ponds. The aim was then to implement fast, precise and reliable mapping methods, making it possible to follow mangrove alterations resulting from these aquaculture activities. Analysis of the radiometric behaviour of the various plant formations of the marsh, based on SIR-C radar data, revealed that the discrimination between these formations is easier when the frequency used is low and the polarization is crossed. Thematic analysis using textural classification was carried out with the L-band and polarization VH. The results generated are promising insofar as the mapping of mangroves is concerned with a fast treatment time and acceptable precision of the results.

Introduction

Mangroves are regarded as major ecosystems due to their intermediate position between the marine and terrestrial environments, their high level of primary productivity (Amarasinghe & Balasubramaniam, 1992) and their nutritional role for coastal shellfish and fish populations (Laegdsgaard & Johnson, 1995; Kimani et al., 1996; Primavera, 1997). Worldwide, they cover an area of nearly 10 million hectares, thus accounting for approximately 1% of the surface covered by tropical forests. The mangroves have relatively strict ecological requirements (e.g. high air temperature, presence of sheltered and relatively shallow areas, wide tidal range, short periods of immersion and preferential presence of muddy substrates; Chapman, 1974). Also, they are commonly unstable and often in regression as a direct result of anthropogenic disturbances, such as those generated by aquaculture activities, which directly or indirectly destroy the canopy (Dierberg & Kiattisimkul, 1996; Binh et al., 1997).

In Madagascar, aquaculture in mangrove areas started rather tardily, with the setting up of a medium-scale shrimp farm along the north-western coast in the beginning of the 1990s (Mahajamba bay). In order to measure the environmental impact of this farm, an assessment of the environmental conditions before the onset of this auqaculture activity (production beginning in October 1993) was considered essential, as was the periodic monitoring of the environment during the duration of this activity. For this type of environmental assessment, the use of remote sensing produces fast, reliable and precise results. The first objective of the present study was, therefore, to determine the environmental conditions of the period preceding the setting up of the aquaculture activity using SPOT-XS data.

Although the data acquisition systems of multispectral data, such as SPOT or LANDSAT, represent suitable tools for mapping mangroves and assessing their overall condition (Dutrieux et al., 1990; Green et al., 1996, 1998; Long & Skewes, 1996; Ramsey & Jensen, 1996), their use in tropical coastal areas is often limited due to an almost permanent cloud cover. Thus, recent research efforts have turned to the use of remote sensing radar (Hery et al., 1993; Mougin et al., in press; Proisy et al., in press), as this image acquisition system is not hindered by cloud cover. Based on SIR-C radar data of the marsh of Mahajamba, the radiometric behaviour of the mangroves, as evaluated from their densities, was examined and compared to other tropical plant formations. Remote sensing of the mangroves, using radar data, is generally carried out using a purely radiometric approach. Thematic analysis using radiometric and textural classifications was also considered here in order to define mangrove mapping methods. The long term goal of this study is to generate a fast and reliable tool to be used in the monitoring of mangrove habitats subjected to anthropogenic activities.

Materials and methods

The study site is located in North-Western Madagascar, between latitudes 15° 10′ S and 15° 45′ S, and longitudes 46° 45′ E and 47° 20′ E (see Figure 1). Mahajamba bay is the point of convergence of two large coastal rivers: the Sofia (in the east) and the Mahajamba (in the south; see Figure 2). Mean annual precipitation is in the order of 1500 mm, occurring for the most part between November and April (Chaperon et al., 1993). Local tidal amplitude is 4–4.50 m during the spring tides.

SPOT-XS imagery (20 m pixel), taken on 21 April 1993 (10 h15 standard time, tide of +2.02 m), and composed of 3 rough channels (XS3, XS2, XS1), was analyzed using GEOimage® software. A low level of cloud cover was present on this date, although some clouds can be observed to the south-east and smoke is perceptible in the centre of the image. A pretreatment was first performed, the purpose of which was to carry out radiometric and geometrical corrections to the rough image. The geometrical correction consists of transferring onto the image, and subsequently onto the reference document (topographical maps to the 1/100 000: N-39, 1953; O-39, 1956; N-38, 1953; O-38, 1953), several reference points; the relative coordinates of which are recorded in each reference mark. These coordinate lists generate a polynomial model, making it possible to pass from the reference mark on the image to that of the map. The projection system used is of the Madagascar Laborde type with international 1967 ellipsoïde. Distinguishing between the immersed and terrestrial areas was made possible using a plant index so as to avoid mistaking slightly immersed areas and mangroves. The terrestrial zones were mapped using supervised classification (maximum probability) with XS3/XS2/Plant Index channels, by successive masks on the mangroves, swamps and terrestrial ecosystems. Manual corrections were carried out based on field data obtained between 1993 and 1995.

SAR radar imagery (Synthetic Aperture Radar) from the satellite ERS-1, type SIR-C (Spaceborne Imaging Radar-C), taken on 7 October 1994 (11 h15 standard time, tide of +0.34 m), presents a portion of the study site, and in particular the mouth of the Sofia and Mahajamba rivers. This multifrequency data was obtained as C-band (5.35 GHz or 6 cm) and L-band (1.25 GHz or 23 cm) with VV and VH polarization (Resolution: 15 m; Width of sweep: 35–100 km; View angle: 15–60°; Length of aerial: 12.1 m). Each image pixel represents the amplitude of the retrodiffused signal. This numerical data (DN) was transformed so as to obtain the retrodiffusion coefficients (dB) for each pixel (dB=−40.2+0.2DN; Pers. comm.: NASA). A filtering was then applied to the images in order to reduce the shimmer phenomenon (speckle). Thematic analysis using radiometric classification was carried out by directed classification (maximum probability), on the basis of both the L-band associated with two polarizations and the C-band associated with crossed polarization, namely L-VH, L-VV and C-VH. Thematic analysis using textural classification was carried out using the L-band associated with cross polarization and the PAPRI software (Borne, 1990; CPIS, 1996). The choice of the L-VH channel is explained by the fact that low frequencies cross natural mediums with more ease, and cross polarization accentuates discrimination between the plant formations. This method makes it possible to threshold the image in a supervised manner, based on the textural heterogeneity of the structures. Image analysis was carried out using a pixel window. For a given window, the closest landscape and the distance (distance between the theoretical composition of the landscape and the real composition of the window) were calculated. Image analysis was carried out on a set range of window

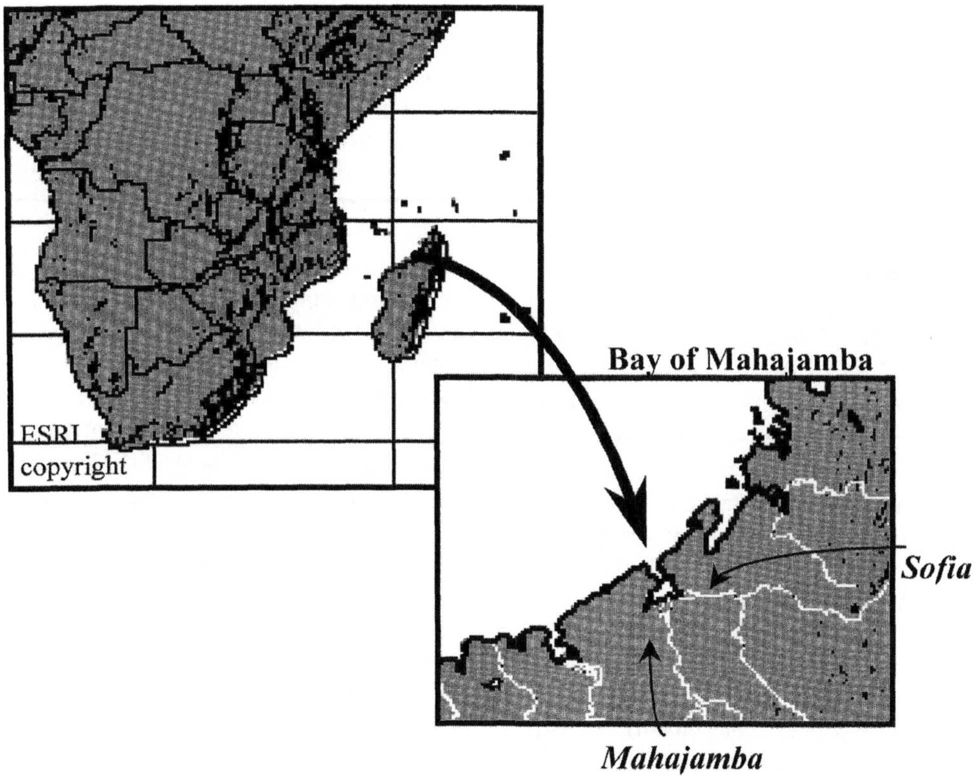

Figure 1. Geographical location of the study site.

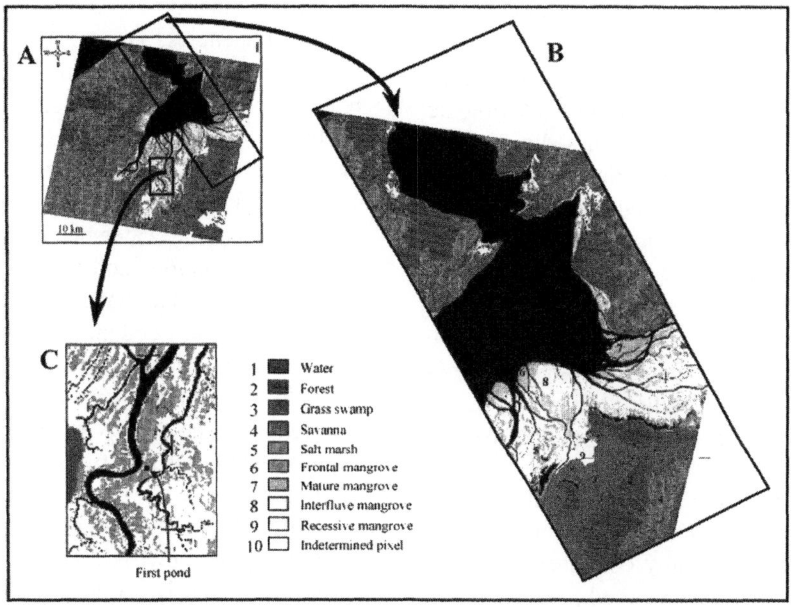

Figure 2. Mapping of the tropical plant formations by image processing of SPOT-XS: (A) of Mahajamba bay, (B) of the area surrounding the shrimp farm and (C) at the mouth of the Mahajamba and Sofia rivers.

sizes, in order to provide the best possible classification of each structure to be mapped. Three tests are realised with 3×3 and 5×5 windows, with 29×29 and 31×31 windows and from 3×3 to 31x31. A geometrical correction was then carried out based on image reference points using the SPOT data and this was done in order to render them strictly superposable.

Results and discussion

The analysis of SPOT-XS imagery made it possible to map Mahajamba bay in April 1993 and revealed the presence of several different environments: mangrove ecosystems, terrestrialecosystems (savannahs and forests) and swampy areas (see Figure 2A). The surface area covered by the coastal marsh is 56 400 ha, of which 43 700 ha is made up of mangroves. These mangroves are well developed in the southern and eastern regions of the bay, at the mouth of the Sofia river and in the western parts. Some small mangrove islands are present to the north. Four types of mangroves have been identified: two types from morphological parameters (frontal, interfluvial) and two other types from dynamic parameters (mature, recessive). This zonation is characteristic of the mangrove ecosystem's dynamics in Madagascar (Lebigre, 1990). The levels of fluvial and sedimentary input determines the spatial evolution of the mangroves. At the mouth of the Sofia river, in the eastern part of the bay, fluvial input is high and the mangrove map shows intensive hydrosedimentary dynamics in this region (see Figure 2B). Large quantities of laterite mud accumulate in the shallow waters and almonds of sediments are formed as one approaches the sea. Pioneer mangroves (e.g. *Sonneratia alba, Avicennia marina*) can then develop, allowing a fast propagation. In the mouth of the Sofia, the mature mangroves (*Rhizophora* spp.) are well developed (see Figure 2B), revealing the intensive dynamics present at this site. A previous study of the spatiotemporal evolution of mangroves in the mouth of the Sofia would tend to confirm this result (Lebigre, 1990). Conversely, in the central part of bay, at the mouth of the Mahajamba, mature mangroves are less developed (see Figure 2B), which would suggest more reduced dynamics. Most of the waters of the Mahajamba basin are artificially contained upstream of the coastal plain (Chaperon et al., 1993). Beyond the mature mangroves and in the region of the interfluvial mangroves, many salt marshes (brine ecosystems without vegetation) develop (see Figure 2C). In April 1993, the first of the shrimp farming ponds was set up in one of these marshes (see Figure 2C). This aquaculture activity currently occupies several hundred hectares, installed primarily on salt marshes and some mangrove areas.

The SPOT-XS data made it possible to characterize the Mahajamba bay environment prior to the development of the aquaculture ponds. These data will be used to monitor the evolution of this region over time. However, the use of SPOT data presents several disadvantages. Firstly, the use of this data is not always possible, due to an almost permanent cloud cover. The size of the pixel (20 m in multispectral mode) also represents a limiting factor in the precise monitoring of mangrove evolution. Lastly, the mapping of tropical areas using SPOT data can lead to a number of confusions and requires that a classification using masks be performed which tends to confer a large amount of subjectivity to the manipulator. The radiometric behaviour of the mangroves, as determined using SIR-C data, has, therefore, been determined based on the density of these ecosystems, which can be compared to the other plant formations present. The variation in mangrove density is easily detectable using SIR-C radar data (Proisy et al., in press). This analysis (see Figure 3), carried out on the four channels available, reveals a similar general tendency in the channel response for the four channels considered. The use of cross polarization, however, accentuates this discrimination between the plant formations at lower frequencies. For the L-band, the dense mangroves are easily identifiable. As for the diffuse and regressive mangroves, they can be distinguished only on the L-VH channel. The C-band would appear to be of interest in differentiating between forests and swampy meadows. It seems that these answers are influenced by the water or humidity present beneath the plant formations and will differ according to the wave length emitted. Savannahs are easily identifiable regardless of the frequency of the signal and even under HH polarization (Proisy et al., in press). Moreover, the value of the retrodiffusion coefficient increases as a function of the density of the plant formations (see Figure 3) and thus as a function of biomass (Proisy et al., in press).

On the basis of these observations, thematic analysis using radiometric classification was carried out using the L-VH, L-VV and C-VH channels (see Figure 4). These results reveal a certain amount of confusion in the interpretation. Indeed, recessive mangroves cannot be distinguished from the forests, nor the dif-

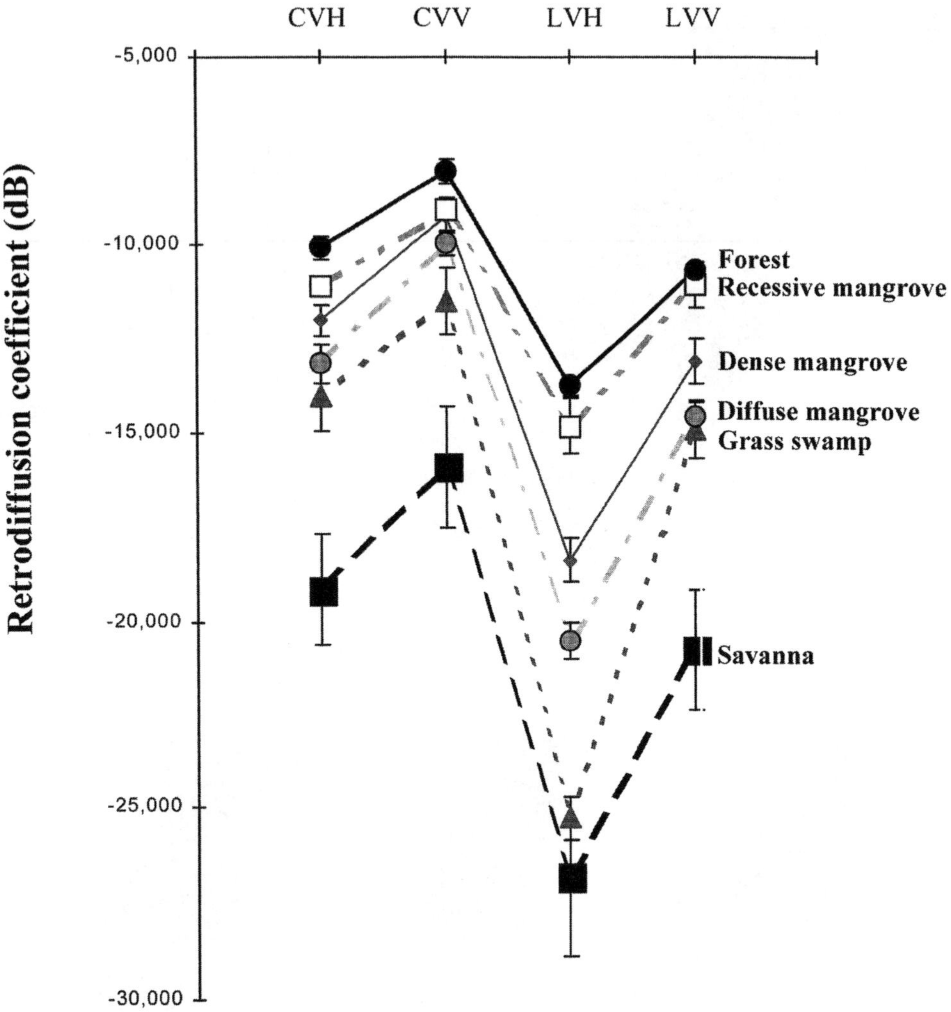

Figure 3. Radar responses concerning the tropical plant formations in the L-band and C-band associated with VH and VV polarizations (Retrodiffusion coefficient depends on the amplitude of retrodiffused signal and is expressed by the relation: retrodiffusion coefficient=$(K^* R4^*\mathrm{Pr})/Sr$; with K=wedging coefficient, R=cible distance, Pr=received power, Sr=area of observed surface).

fuse mangroves from savannahs, the salt marshes from swampy meadows and water from swampy meadows. Thematic analysis using radiometric classification of the SIR-C radar data, therefore, does not appear to be a reliable method for precisely mapping mangroves and other tropical plant formations. Thematic analysis using textural classification was thus carried out using only the L-VH channel (see Figure 5). The test carried out, with 3×3 and 5×5 windows (see Figure 5A), provides interesting results in terms of the identification of dense mangroves, water and swampy meadows, with values in the confusion matrix higher than 84%. Using 29×29 and 31×31 windows (see Figure 5B), most of the structures can be identified, with the exception of forests and salt marshes. As the salt marshes are totally confused with water, it is assumed they have the same texture. The forest is, in places, confused with the recessive mangroves; they thus have the same texture and, consequently, the same density. Moreover, the delimitations of the different plant formations are not very distinct (see Figure 5B). Consequently, a third test is considered using a larger window from 3×3 to 31×31 (see Figure 5C). This treatment allows us to more clearly distinguish the delimitations of the plant formations, but does not reduce the overlapping of salt marshes and forests. Thematic analysis using textural classification of SIR-C radar images, therefore, generates promising results in terms of the precise mapping of mangroves and other tropical plant formations, with the exception of forests and salt marshes.

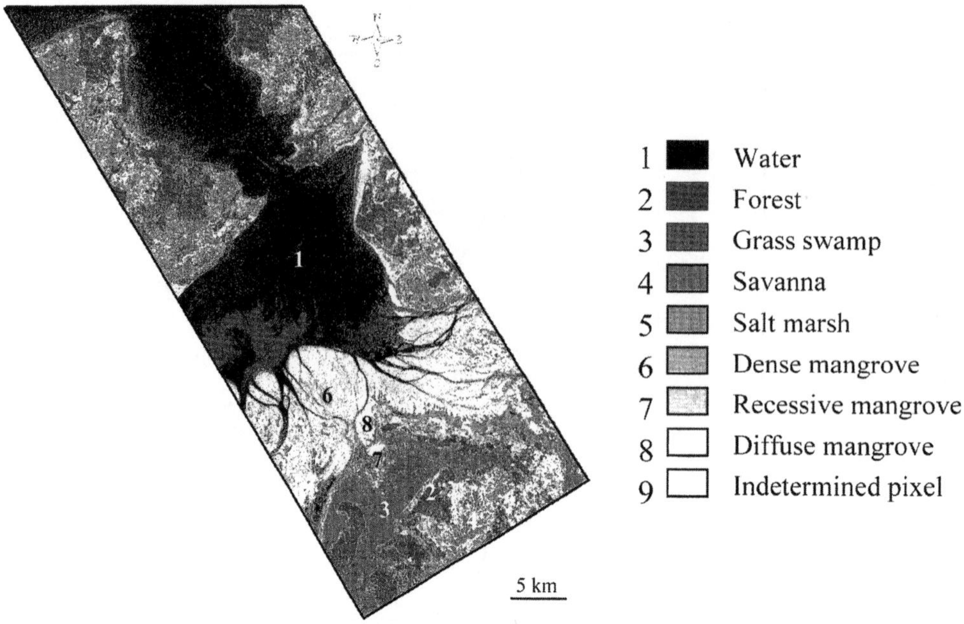

Figure 4. Mapping of the tropical plant formations in the mouth of the Mahajamba and Sofia rivers using radiometric classification of SIR-C radar images.

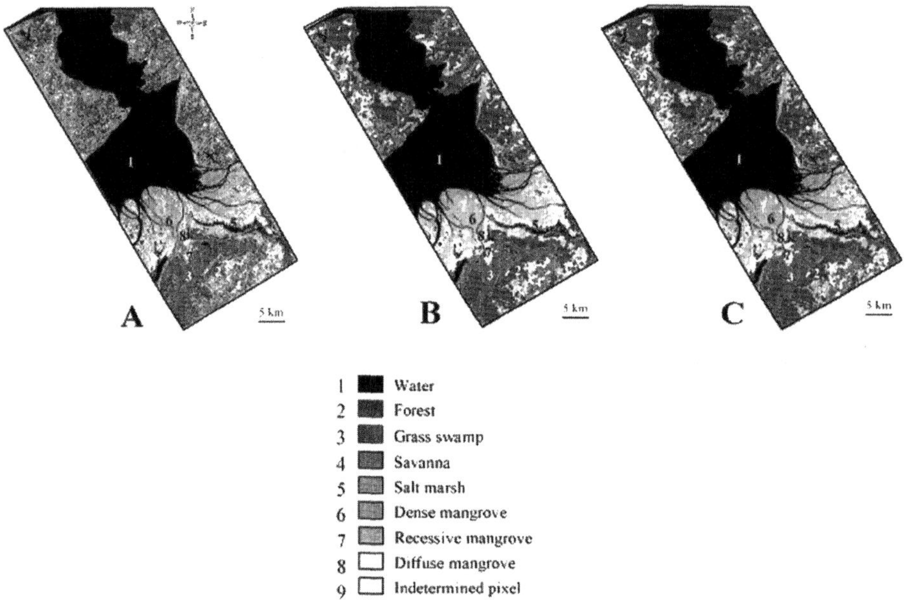

Figure 5. Mapping of the tropical plant formations in the mouth of the Mahajamba and Sofia rivers using textural classification of SIR-C radar images: (A) with 3×3 and 5×5 windows, (B) with 29×29 and 31×31 windows, (C) from 3×3 to 31×31.

Conclusions

The mapping of Mahajamba bay, using SPOT-XS data, made it possible to characterize the evolutionary dynamics of the mangroves present within the bay prior to the setting up of aquaculture rearing ponds. The mouth of the Sofia river is characterized by intensive dynamic conditions in its frontal area. Conversely, the central part of the bay exhibits a more reduced evolution. Analysis of the radiometric behaviour of the mangroves and other plant formations in Mahajamba bay, as determined from SIR-C radar data, reveals that these formations are more easily identified when the frequency used is low and when polarization is crossed. Thematic mapping using radiometric classification did not generate satisfactory results. Conversely, textural classification using the L-VH channel provided interesting results, insofar as the precise mapping of mangroves and other tropical plant formations is concerned. Results for forests and salt marshes using this technique, however, were not satisfactory. In order to monitor the environmental evolution of mangroves in relation to the development of aquaculture activities in Mahajamba bay, the textural analysis of SIR-C radar images appears to be the technique most adapted to our objectives, with the added advantage of possessing a short treatment time.

Acknowledgements

We make a point of thanking Christian Reynaud, researchers at the Laboratory of Marine and Continental Hydrobiology in Montpellier (France), for the interest they showed in this work. Our thanks also go to Frederic Borne, researcher at the CIRAD Laboratory, Centre for Remote Sensing (Montpellier) and André Beaudoin, researcher at the CEMAGREF Laboratory, Centre for Remote Sensing, for their valuable advice and help in the completion of this study.

References

Amarasinghe, M. D. & S. Balasubramaniam, 1992. Net primary productivity of two mangrove forest stands on the northwestern coast of Sri-Lanka. Hydrobiologia 247: 37–47.

Binh, C. T., M. J. Phillips & H. Demaine, 1997. Integrated shrimp-mangrove farming systems in the Mekong delta of Vietnam. Aquacult. Res. 28: 599–610.

Borne, F., 1990. Méthodes numériques de reconnaissance de paysages. Thèse de 3ème cycle, Université Paris VII: 213 pp.

Chaperon, P., J. Danloux & L. Ferry, 1993. Fleuves et rivières de Madagascar. ORSTOM Paris: 874pp.

Chapman, V. J., 1974. Mangrove biogeography. In Walsh G., S. Snedaker & H. Teas (eds), Proc. Inter. Symp. Biol. Management of Mangroves, Honolulu, Hawaï, U.S.A: 3–22.

CPIS, 1996. Classement Paysager d'Images Satellites; Guide de l'utilisateur. CIRAD-INRA (eds): 32pp.

Dierberg, F. E. & W. Kiattisimkul, 1996. Issues, impacts and implications of shrimp aquaculture in Thailand. Envir. Mgmt 20: 649–666.

Dutrieux, E., J. Denis & J. Populus, 1990. Application of SPOT data to a base line ecological study of the Mahakam delta mangroves (East Kalimantan, Indonesia). Oceanol. Acta 13: 317–326.

Green, E. P., C. D. Clark, P. J. Mumby & A. C Ellis, 1998. Remote sensing techniques for mangrove mapping. Int. J. Remote Sensing 19: 935–956.

Green, E. P., P. J. Mumby, A. J. Edwards & C. D. Clark, 1996. A review of remote sensing for the assessment and management of tropical coastal resources. Coast. Mgmt 24: 1–40.

Hery, P., D. Ducrot-Gambart, A. Lopes, E. Mougin, G. Marty, F. Fromard, F. Blasco, J. P. Rudant, M. Lointier & M. T. Prost, 1993. Cartographie de la végétation tropicale et suivi de la dynamique des mangroves de Guyane par SPOT et ERS-1. In Cépaduès eds, De l'optique au radar, les applications de SPOT et ERS, Mai 1993, Paris, France: 329–341.

Kimani, E. N., G. K. Mwatha, E. O. Wakwabi, J. M. Ntiba & B. K. Okoth, 1996. Fishes of a shallow tropical mangrove estuary, Gazi, Kenya. Mar. Freshwat. Res. 47: 857–868.

Laegdsgaard, P. & C. R. Johnson, 1995. Mangrove habitats as nurseries: Unique assemblages of juvenile fish in subtropical mangroves in eastern Australia. Mar. Ecol. Progr. Ser. 126, 1–3: 67–81.

Lebigre, J. M., 1990. Les marais maritimes du Gabon et de Madagascar. Thèse doct. Géographie, Univ. Bordeaux III: 704pp.

Long, B. G. & T. D. Skewes, 1996. A technique for mapping mangroves with Landsat TM satellite data and Geographic Information System. Estuar. coast. shelf. Sci. 43: 373–381.

Mougin, E., C. Proisy, G. Marty, F. Fromard, H. Puig, J. L. Betoulle & J. P. Rudant, 1999. Multifrequency and multipolarization radar backscattering from mangrove forests. IEEE Transact. Geosci. Remote Sensing 37(1): 94–102.

Primavera, J. H., 1997. Fish predation on mangrove-associated penaeids – The role of structures and substrate. J. Exp. Mar. Biol. Ecol. 215: 205–216.

Proisy, C., E. Mougin, F. Fromard & J. P. Rudant, 1998. Télédétection radar des mangroves de Guyane Française. Photo-interprét., Images spatiales et aériennes, num. Spéc. in press.

Ramsey, E. W. & J. R. Jensen, 1996. Remote sensing of mangrove wetlands: relating canopy spectra to site-specific data. Photogramm. Engin. Remote Sensing 62: 939–948.

Hydrobiologia **413:** 135–142, 1999.
R.S. Dodd (ed.), Diversity and Function in Mangrove Ecosystems.

The mangrove: an area to be cleared... for social scientists

Marie-Christine Cormier-Salem
IRD (ex-Orstom)/MNHN, Laboratoire d'Ethnobiologie–Biogéographie, 57 rue Cuvier - 75231 Paris cedex 05
Tel: [+33]-1-40-79-36-63. Fax: [+33]-1-40-79-38-15.
E-mail: cormier@mnhn.fr

Key words: integrated approach, coastal wetlands, man-nature relationships, actors, representation, management

Abstract

In the last two decades, fields of research concerned with the mangrove have been enlarged. Analysis methods and tools have been renewed. Mangroves losses and the desire for restoration have stimulated the development of integrated approaches. These approaches – comprehensive, multidisciplinary or whole-system – tend to take into account not only naturalists' perpectives, but also those of social scientists and managers'. The purpose of this paper is to examine, firstly, the changes in the scientific representation of mangroves; secondly, the place of social sciences in research programs. Information stems on the one hand from a broad review of the literature and, on the other hand, from fieldwork, multidisciplinary workshops and publications about the West-African mangrove coast, conducted in the framework of DUM (**D**ynamic and **U**ses of **M**angrove) program of Orstom (the French National Institute of Research for Development in Cooperation). This paper argues that, in spite of ecological awareness and widening knowledge, coastal marshes are always mis-understood. Up to now, social scientists' perspectives are not considered enough, although natural scientists have called for their participation in an integrated approach. More attention should be paid particularly to the relations between mangroves and men, and thereby to the co-evolution of mangrove natural and social systems, so that long-term management can be effective.

Introduction: changes in mangrove representation

The mangrove was for a long time considered as a waste, sterile, unhealthy marsh; then, in the 1970s, as a valuable ecosystem; and, in the last ten years, as a complex space under multiple pressures (Cormier-Salem, 1994b). According to these changes in our perspectives, mangrove studies have been improved (see Figure 1). Fields of research have been enlarged[1]. In fact, our interest in the mangrove began with botanists, collaborating with ecologists, and, lastly, with social scientists.

Now, more and more research projects, concerned with intertidal marshes, point out the desire to take into account socio-economic perpectives. Most of the decision-makers agree that effective management depends on whole-system understanding. Besides, research tools have become more and more sophisticated, thanks to remote sensing, spatial analysis, GIS etc.

In spite of ecological awareness and widening investigation, mangrove coastal marshes are still mis-understood, and therefore mis-managed. This paper has for its purpose to underline the gaps between scientific knowledge and management projects. Information is based on a broad review of the literature, fieldwork, multidisciplinary workshops and publications conducted in the framework of DUM program (Cormier-Salem, 1994b, 1999). This program '**D**ynamic and **U**ses of **M**angrove' (1993–1996), coordinated by Orstom (the French National Institute of Research for Development in Cooperation became, in January 1999, IRD Institute of Research for Development), has been centered on the West-African coast from the Saloum region (Senegal) to Sierra Leone, called in French '*Rivières du Sud*', in English '*Northern Rivers*' or '*Upper Guinea Coast*' (see Figure 2). The main goals of this program were, firstly, to demonstrate the spatial and temporal diversity in

[1] In the frame of this paper, it is impossible to quote all the synthesis and literature reviews about mangroves. For more complete references, see: Cormier-Salem, 1999, volume 2.

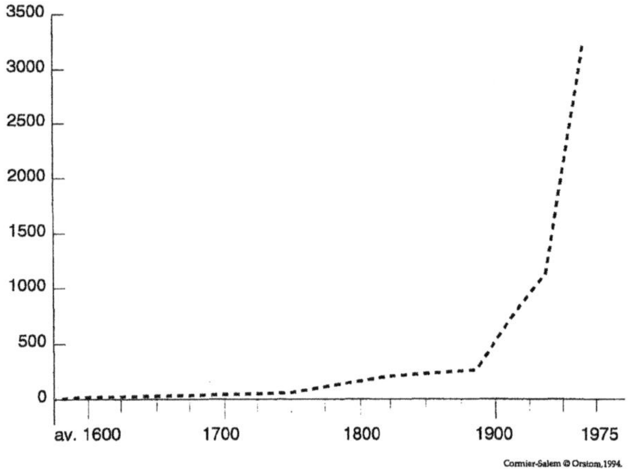

Figure 1. Progress in Knowledge of Mangroves according to the publications from 1600 to 1980 (from Rollet, 1981).

the uses and values of mangroves; secondly, to examine the 'state of the art' in our understanding of the *Northern Rivers* mangrove; thirdly, to question our pre-conception of coastal wetlands (healthy or unhealthy, rich or poor, fragile or robust, virgin or overcrowded) using a multidisciplinary and comparative approach. In addition, an annotated review of the references about the West-African mangrove coast has been published. Lastly, a network of mangrove specialists from the natural and social sciences has been set up.

Two aspects will be developed in more detail: at a regional level (*Northern Rivers*), mangrove specificities will be pointed out. At a more global level, mangrove complexities will be examined; research and policies issues will be suggested.

The *Northern Rivers* specificities

In West Africa, managers failed – and are still failing – to meet their goals in managing mangroves, because they have not taken into account enough the ecological and social particularities of mangrove systems.

Ecological particularities

The most striking physical features of the *Northern Rivers* are the spread of coastal wetlands and the numerous rivers – more than two dozen rivers reach the sea – that give the name to this coastal region. As described by Rodney (1970), coastal environment is dominated by a "complex of numerous rivers, low plains, marshy land and powerful tides: a complex which decisively affected all activity conducted on the Upper Guinea Coast". Coastal salt marshes, estuaries – more precisely 'ria' – wetlands and floodplains are covered with mangrove forests, for over 1000 km from the Saloum Islands in Senegal to the Sherbro Island in Sierra Leone (see Figure 2). The inland extent of this amphibuous vegetation varies from a few kilometers to more than 150 km. From Diop (1990), the *Northern Rivers* mangroves *stricto sensu*, that means 'existing mangrove communities', cover 1 000 000 ha; but, *largo sensu*, that includes inland mangrove areas (very often reclaimed for rice cutivation) and continental shelf, the surfaces reach 3 000 000 to 3 500 000 ha.

The vegetation, almost the same all over the region, consists of nine species distributed in four families (Sow et al., 1994). Rhizophorae are dominant with three main species: *R. harrisonii*, *R. mangle* and *R. racemosa*. Avicenniaceae are represented by one species, *Avicennia africana* (or *A. nitida* in Sierra Leone). Combretaceae are represented by two species: *Conocarpus erectus* and *Laguncularia racemosa*. Malpighiacea are also abundant (notably in Guinea) with three main species: *Banisteria leona*, *Heteropteris leona* and *Stigmaphyllum ovatum*.

Due to rainfall reduction since the end of the 1960s, vegetation distribution and composition have been modified. Today, the prevalent vegetation consists of *Avicennia*. However, the ecological dynamic of the *Northern Rivers* mangroves is contrasted. Two areas can be delineated (Cormier-Salem, 1994b):

1. From Saloum (Senegal) to Geba (in Guinea Bissau), mangrove ecosystems are retreating and becoming less rich because of drought, salinisation and acidification. Water shortage entailed an increased salinity, up to 170 ppt in the middle part of the Casamance River in June 1986 ! The rapid and drastic ecological degradation has modified all the components of the trophic system: *Rhizophora* tend to disappear; *Avicennia*, more resistant to salinity, took their place but they are also declining; the hypersaline and barren surfaces, called *tanne*, are expanding more and more. Since 1969, 70–80% of mangroves have been lost (Marius, 1995). Moreover, ichtyofauna and avifauna are becoming less diversified. Fish community sizes, abundance and species number have been diminishing.
2. From Geba to Sierra Leone River, mangroves have a remarkable dynamic of progression and regression. In Guinea (Ruë, 1998) as in Sierra Leone (Anthony, 1990), mangrove forests are either retreating or advancing, according to hydro-climatic environments. Also, the Guinean coast advanced forward 80 m per

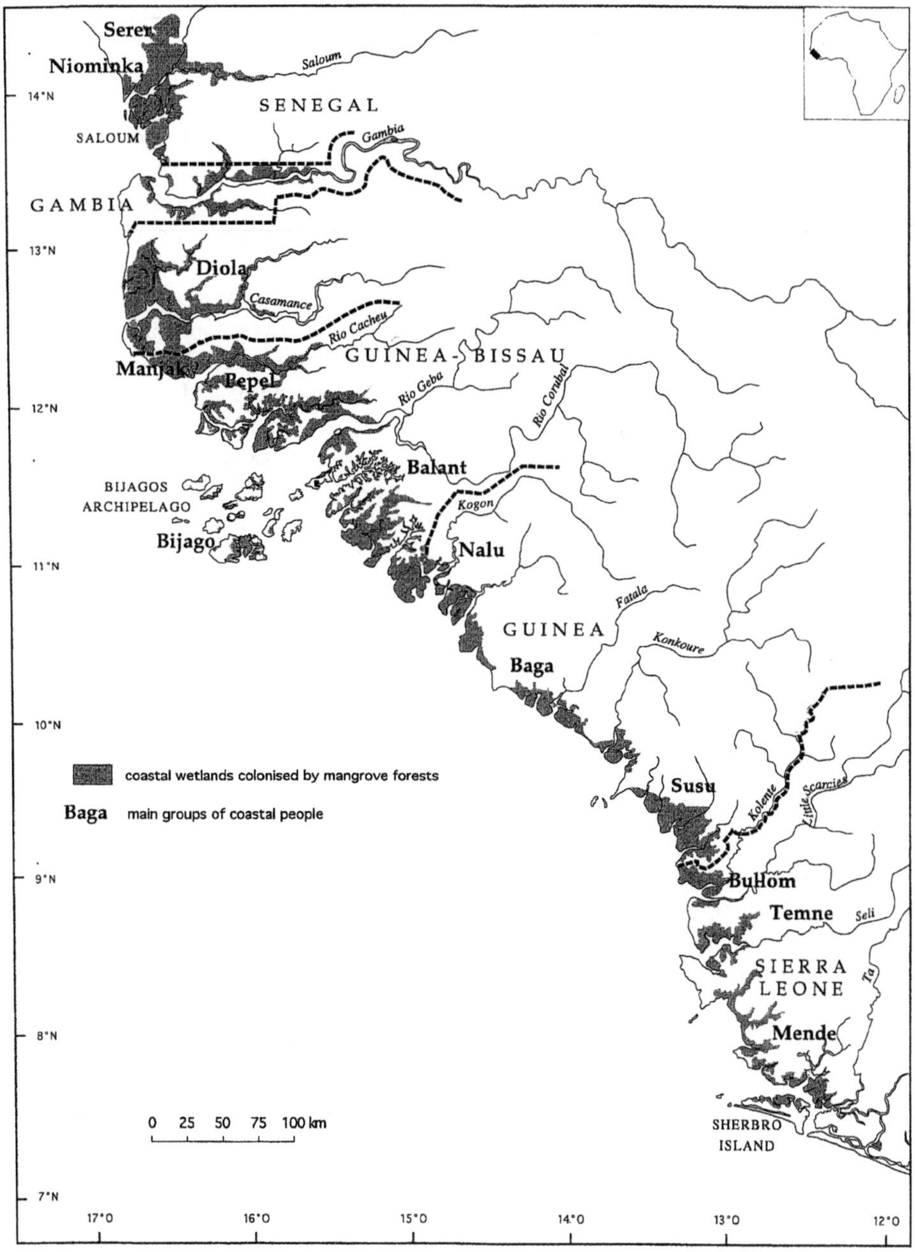

Cormier-Salem © Orstom, 1998

Figure 2. Northern Rivers (West Africa): Land and People.

year between 1958 and 1965, but retreated 100 m per year between 1983 and 1987 (Cormier-Salem, 1999). Rainfall (more than 4000 mm in this area, *versus* less than 1500 mm in the Saloum-Geba area) and wind cycles seem to be the major factors of this hydrosedimentary dynamic of the mangrove. The processes of coastline erosion and accretion, mangrove regression and progradation are periodic and reversible.

Social particularities

The most striking social feature of the *Northern Rivers* is the strong relationship between mangroves and local people. For a very long time, mangrove swamps have been occupied, managed and controlled by cultivator-fishermen groups, such as Diola, Baga, Manjak, Nalu etc. (see Figure 2). These people belong to an old agrarian civilisation, based on rice cultivation. They form egalitarian peasant societies, without classes, or state superstructure. The family is the sole effective social and political unit. The majority of them are animist. They differ radically from the hinterland people (Mande or Foulah), islamised and characterised by slavery institutions and hierarchical structures, with classes of landlords, horsemen warriors and marabouts.

The old intimacy of the *Northern Rivers* communities with mangroves is expressed through the social, cultural, economic and political institutions, the high densities of population and the construction of amphibuous territories (see Plate 1). Mangroves have been reclaimed for rice farming and used for multiple natural renewable resources (timber, firewood, charcoal, tanin, salt, fish, crabs, oysters, shrimps, etc.) at least since the 15th century (Cormier-Salem, 1992, 1999). The Portuguese navigators, who discovered the West-African Coast in the mid-15th century, have given the first written sources of information about the *Northern Rivers* (or *Guinea*) population and land. Before the arrival of the Portuguese, archeological and linguistic sources (Linares, 1971; Bühnen, 1992) give evidence of a very old rice farming system, as old as that of the Niger Central Delta, based on an original African wet rice species (*Oryza glaberrima*).

Mangrove swamp rice cultivation requires an intensive and sophisticated farming system, without an equivalent in any other country: soils under mangroves are salty and potentially acid sulphate. Material is muddy, heavy and sometimes very liquid. These environmental constraints are overcome, thanks to a remarkable control of water (salty marine water and fresh rainfall water) and a huge investment in manual labour. The clearing of mangroves, construction of dams and controlling of tides are the basic operations of the flooded rice cultivation. These operations are realized with a specific ploughing tool, called *kajendu* among the Diola speakers and used all along the coast.

The continuity of this 'mangrove system' up to now is amazing. Nevertheless, continuity does not mean uniformity in space and/or time. Components of this system have changed according to environnemental modifications, the strategies of actors and the processes of innovation (Cormier-Salem, 1994a). In the last few decades, the salinisation of soil and water, attributed to rainfall reduction, has produced a mangrove rice cultivation retreat. But, climatic pejoration has accelerated and emphasized a process, begun in the 1950s with urbanisation and migration toward cities. Management shortage, due to the young workers departure, appears to be the first factor of the degradation of the mangrove system. Rural exodus appears as the first factor in time, but perhaps also in importance. The farmers have to develop innovative strategies to sustain this system, such as new combinations of resources, uses and techniques.

Also, three main trends are listed: First, the retreat of salty swamps rice cultivation is compensated by the colonisation of new lowlands or uplands, exploited for wet and/or dry rice fields, but also for diverse corn, fruit and vegetable cultures. Next, the collection of aquatic and terrestrial resources within mangroves is becoming more and more developed. Natural resources – such as oysters, salt, timber, shrimps etc. – are now considered as market products, providing an essential source of income for local communities. Lastly, new productive activities have emerged, as the small-scale marine fishery. Therefore, cultivator-water inland fishermen groups have to adopt new techniques – such as large nets and motorized canoes – undertake marine migrations and set up fish distribution.

Finally, the *Northern Rivers* people have manifested their ability to face environmental changes, thanks to the diversity of mangrove resources. The mangrove, as a multiple use system, appears flexible.

Among the main results of the DUM studies on the *Northern Rivers*, the mangrove complex – both an ecosystem and a multiple use system, e.g. a natural and human system – should be considered sensitive, rather than fragile. Instability and process reversibility (or resilience) rather than irreversibility and crisis should be underlined (Cormier-Salem, 1996).

The mangrove under debate: research and policy issues

In regard to these diverse and remarkable dynamics, what is the most convenient management? What should the aims be of the mangrove management projects: exploitation, conservation, protection, restoration?

Failures of West-African mangrove management

Economic valorisation of mangrove is a way to increase primary production, improve national consumption and also gain foreign currencies through high values products export, as shrimp. In West Africa, conversion of marshes to rice paddies was the aim of colonial administrators and is still the main purpose

Plate 1. Mangrove communal territory (©Cormier).

of technical experts and managers. But most of these large scale and intensive projects failed.

During the Colonial and immediate post-Independance period, a model from Indochinese deltaic lowlands management, was transferred to the Guinean coastal wetlands, although environmental conditions differ (Ruë, 1998). In Guinean coastal areas, rainfall is abundant (more than 4000 mm) but limited to a short period of the year. Drying and salinisation through marine inflow (from November to April) and fresh water submersion (between May and October) are major constraints, besides the fluidity of the soils and the heterogeneity and instability of the lands. In contrast, in the Mekong Delta, fresh water supply, from rainfall and fluvial inflow, is abundant and regular all year round. Soils, not submitted to tides, are not salty. Lowland surfaces are flat and relatively homogeneous. Finally, very high population densities and the use of buffalo have made the work of cultivating rice easier. Indochinese hydraulic management was poorly adapted both to social and ecological conditions. European managers did not take into account traditional knowledge and were unaware of local conditions and skills. Rural communities have been marginalized. Reclamation of mangrove marshes were, for the most part, ineffective for increased rice production, and at the same time led to mangrove losses, because of natural dynamic perturbations. Another example, more recent (end of the 1970s), is found on the north shore of Casamance River: hydraulic strategies, transferred from the Netherlands, consisted of wetlands drainage and polderisation processes. They led to irreversible acidification of vast surfaces, that are now barren and sterile. Specialists of mangrove soils (Marius, 1985, 1995; Sow et al., 1994) reveal the negative effects of coastal wetlands drainage, especially during drought period. They assume wetlands under salt-water (marine and tidal) inflows are less altered than without water. On one hand, salty surfaces could be colonized again by vegetation, as soon as rainfall is sufficient; on the other hand, drying surfaces become ever too acidic for any vegetation.

Therefore, rice fields, as mangroves forests, have disappeared. Mangrove conversion for one single use makes systems more simple, and then more vulnerable (Cormier-Salem, 1992, 1996).

Multiple values and functions of mangroves lead to the setting up of restoration projects and natural reserves. Nevertheless, what is supposed to be conserved or restored: forests, fish or bird populations, aquatic ecosystems, or mangrove settlements and communal territories? In other words, what is the view of the state for a mangrove definition?

Diversity of scientific definitions of mangroves

Scientific conceptions of mangroves depend on the domain and purpose of research. For a botanist, at a basic level (micro-scale), the mangrove is usually defined as an evergreen tree species, most often of the Rhizophorae or Avicenniaceae family, growing in a zone of tidal influence. At a larger level, it is considered as a complex of plant communities and could be referred to mangal, as Macnae (1968) suggested. In fact, the two terms, 'mangle' and 'mangrove' are equally used (in French, the sole term 'mangrove' is used), and the context is supposed to make the meaning of the used term clear. Nevertheless, the term is not always explicitely defined. This ambiguity produces scientific bias. Also, the number of 'mangrove' species, cited by different sources, varies from 53 to 75 (FAO, 1994).

Moreover, the delimitation of 'mangrove' areas could vary significantly, according to the various scientific conceptions. For an ecologist, whose purpose is to understand the function of aquatic ecosystems, all the surfaces under the mangrove influence should be considered, from the continental shelf to intertidal marshes, with or without mangrove trees.

For social scientists, whose purpose is to reveal the multiple uses, roles and values of coastal wetlands, to identify social actors and understand their strategies, observation levels could be restricted to mangrove ecosystems (for instance, to list the use of resources) or broadened to the nearby environment (for instance, to point out mangrove roles among local users, in comparison with upland usage). Larger levels should be also considered: national levels, in order to discuss the efficiency of coastal wetlands legislation and management policies, and global levels, in order to reveal international ecological or economic stakes concerned with mangrove management.

Indeed, various definitions of mangroves (from roots to coastal wetlands and far beyond) are chosen, depending on the specific targets in question, research frame and local context. Deficiencies in standardized criteria make the evaluation of the state (in quality and in quantity) of mangroves very uncertain (Cormier-Salem, 1996). In the case of the *Northern Rivers*, assessments of mangrove areas vary considerably, according to their definition, besides the different method of measurement and mangrove destruction (see Table 1).

Mangroves appear difficult to analyse because of data heterogeneity and complexity. Some researchers underline the fieldwork hard conditions, because of mosquitoes, marsh-peaty soils, mangroves embedded roots etc. Note that mangrove was described as the 'White man's grave' during the European penetration and Colonial period. Maybe, this perception is not out-to-date! Mangrove resources, both aquatic and terrestrial, are multiple, unstable, mobile and, as other renewable natural resources, owned in common. Mangrove resource access is particularly under debate (Hamilton & Snedaker, 1984; Cormier-Salem, 1995). Governemental regulations are ineffective. Customary rules and communities rights are not officially recognized. In addition, values associated with mangrove are extremely diverse and cannot be estimated with usual economic methods. Wetland regulation and valuation processes are controversial (Barbier, 1993; Coultas & Hsieh, 1997). So, mangrove definition is not an academic debate: it has political and social implications.

Mangroves under pressure

Besides the various scientific concepts of mangroves, users' definitions have to be examined. These definitions reveal the complex and evolutive relations between social actors and their marine and coastal environment. In the *Northern Rivers* (Cormier-Salem, 1995, 1998), from the viewpoint of rice cultivator-fishermen groups (Diola, Baga, Nalu etc.), mangrove is the base for their multiple use system, the framework and support of their communal territories, the roots of their way-of-life. From the viewpoint of marine migratory fishermen, mangrove provides lucrative products (fish but also salt and firewood, used to process fish) and fixed sites for their seasonal settlements. As for Foulah pastoralists, pushed during the dry season (from September to May) to coastal marshes in search of salt and grazing ground for their cattle, mangrove is a particularly attractive area, that structures their migration patterns. Local users' perspectives differ from that of private foreign companies, such as intensive fish farming companies, who consider only reclaimed mangrove. They differ also from WWF, IUCN or BIROE perspectives: international institutions aim to maintain the ecological functions of wetlands (basically, habitat for fish and waterfowl

Table 1. Assessment of Northern Rivers mangrove areas (thousand ha)

Gambie	Sénégal	Guinée Bissau	Guinée	Sierra Leone
67 (1980′) (1)	**169** (1980′) (1)	**476** (1970′) (2)	**260** (1980′) (5)	**284** (1970′) (7)
	420 (1990′) (2)	**347** (1980′) (5)	**260** (1980′) (6)	**172** (1980′) (8)
		249 (1980′) (2)	**385** (1990′) (2)	
59 (1990′) (9)	**185** (1990′) (9)	**249** (1990′) (9)	**296** (1990′) (9)	**184** (1990′) (9)
	Casamance			
	93 (1973′) (3)			
	91 (1979′) (3)			
	89 (1983′) (4)			
	250 (1990′) (2)			

[Mangrove+mangrove associated areas]
sources: (1) Saenger et al., 1983; (2) Diop, ed., 1993; (3) Sall, 1980; (4) Badiane, 1986; (5) CCE-SECA-CML, 1987; (6) Ruë,1989; (7) FAO, 1979; (8) Chong, 1989; (9) Saenger & Bellan, 1995.

bird populations and buffer against storm and waves) and consider mangrove as a vulnerable ecosystem or ecotone (Williams, 1990; Dugan, 1993; Coultas & Hsieh, 1997).

Actors, who intend to manage mangrove according to their own interest, have often opposite ideas concerning mangrove (Cormier-Salem, 1997). As a result, conflicts may occur at various levels: at a local level, mangrove could be considered as a portion of territory, whose control is in the hands of coastal settlers. Competition for resource access emerges frequently between autochtonous peoples and outsiders. However, at a national level, mangrove is mainly considered as a virgin area, open to access or a public property, that should be exploited to provide money. Then, the local peoples concerned may be marginalized. Finally, at a global level, according to the Ramsar Convention, mangrove should be protected against human pressures and registered as a natural reserve. Thereby, all the users (settlers or outsiders, sedentary or migratory, private or public) may be excluded.

To prevent conflicts between the people involved and mangrove degradation, actors have to be identified and their strategies understood. Short-term and long-term impacts of conservation or exploitation policies have to be underlined. Multiple and contradictory stakes, related to mangrove management, need to set up consultation forums.

Conclusion

To sum up, three mangrove definitions seem dominant: firstly, from hygienic and productivist standpoints, mangrove is perceived as a sterile, unhealthy malarial swamp, that should be drained and reclaimed. It is defined as a virgin area or a pioneer front. Resources are considered free of access. New intensive speculations, as shrimp farming, are more and more in favour. Secondly, from social and cultural standpoints, mangrove is perceived as a portion of communal territory, support of a multiple use system, that should be sustained. Resources are owned in common, e.g. mangrove access and management are under the control of local groups. Thirdly, from aesthetical and ecological standpoints, mangrove is perceived as a rich and fragile ecosystem, whose conservation is essential for biodiversity concerns. Therefore, it should be protected against human activities. Resources are under national or international control. Some sites, of international importance, are registered in the Ramsar Convention and defined as natural parks or biosphere reserves.

Mangrove conceptions are various and controversial because of the multiple uses and values associated with this system. The 'mangrove actors' (e.g. the wide range of agents who are concerned with the mangrove) are local users (rice cultivators, fishermen, timber or salt collecters etc.) as well as private firms, public agencies and, also, researchers. Following the spirit of Rio, the need to conciliate environment and development, in other words, the need for concertation necessitates active multidisciplinary programs about the co-evolution of ecological and social systems.

However, the place of social sciences, in such program, should be reviewed. Three different viewpoints make social scientists' concerns misunderstood: demands for their input are generally limited to economic evaluation or assessment of the human impact. Next, their participation is required most often in an urgency or crisis context, to resolve conflictual situations or give recipes to stop degradation. Lastly, changing human factors or social variables, that explain other parameters, are not fully taken into consideration. Real cross-disciplinary approaches are supposed to develop conceptual and methodological frameworks, explicitly and together – as natural scientists as social scientists – by the start of the program. So, effective long-term management is to be expected.

References

Anthony, E. J., 1990. Environnement, géomorphologie et dynamique sédimentaire des côtes alluviales de la Sierra Leone. Rev. d'Analyse Spatiale Quantitative et Appliquée, Nice: 27 (28), 187 pp.

Barbier, E. B., 1993. Sustainable use of wetlands. Valuing tropical wetland benefits: economic methodologies and applications. Geograph. J. 159: 22–32.

Bühnen, S., 1992. Place names as an historical source: an introduction with examples from southern Senegambia and Germany. History in Africa 19: 45–101.

Cormier-Salem, M.-C., 1992. Gestion et évolution des espaces aquatiques: la Casamance, Paris, Orstom, coll. Études et Thèses: 584 pp.

Cormier-Salem, M.-C., 1994a. Environmental Changes, Agricultural Crisis and Small-Scale Fishing Development in the Casamance Region, Senegal. Ocean and Coastal Management 24: 109-0124.

Cormier-Salem, M.-C. (ed.), 1994b. Dynamique et usages de la mangrove dans les Pays des Rivières du Sud (du Sénégal à la Sierra Leone), Paris, Orstom, coll. Colloques et Séminaires. 353 pp.

Cormier-Salem, M.-C., 1995. Paysans-pêcheurs du terroir et marins-pêcheurs du parcours. Les géographes et l'espace aquatique. L'Espace géographique 1: 46–59.

Cormier-Salem, M.-C., 1996. ≪Les littoraux à mangrove, des régions fragiles? Eléments de réflexion à partir des conclusions du programme DUM≫, Contribution au colloque de clôture de DURR. Dynamique et Usage des Ressources Renouvelables, ORSTOM, Orléans, 16–17 octobre 1996: 10 pp.

Cormier-Salem, M.-C., 1998. ≪Les eaux troubles des mangroves. Usages multiples, images contrastées des Rivières du Sud (Afrique de l'Ouest)≫ In Barraqué B., N. Vernazza-Licht et al., L'homme et la Lagune. De L'espace Naturel à L'espace Urbanisé. Marseille, édition du Bergier, Travaux de la Société d'Ecologie Humaine: 321 pp.

Cormier-Salem, M.-C. (ed.), 1999. Les Rivières du Sud. Sociétés et mangroves ouest-africaines. Paris, Orstom, 2 volumes (Volume 1: 416 pp.; Volume 2: 288 pp).

Coultas, C. L. & Y. P. Hsieh (eds.), 1997. Ecology and Management of Tidal Marshes. A Model from the Gulf of Mexico. Delray Beach (Florida), St Lucie Press: 355 pp.

Diop, E. S., 1990. La côte ouest-africaine. Du Saloum (Sénégal) à la Mellacorée (Rep. de Guinée), Paris, Orstom, coll. Études et Thèses: 380 pp.

Dugan, P. (ed.), 1993. Wetlands in Danger. Londres, A Mitchell Beazley World Conservation Atlas, in association with IUCN-The World Conservation Union. 187 pp.+index.

FAO, 1994. Mangrove forest management guidelines. Rome, FAO Forestry Paper 117, 319 pp.

Hamilton, L. S. & S. C. Snedaker (eds.), 1984. Handbook for Mangrove Area Management. UNEP/UNESCO/EAST-WEST Center/IUCN.

Linares, O. De Sapir, 1971. Shell middens of lower Casamance and problems of diola protohistory. W. Afr. J. Archeol. 1: 23–54.

Macnae, 1968. A general account of the Fauna and Flora of mangrove swamps and forests in the Indo-West-Pacific region. Adv. Mar. Biol. 6: 73–269.

Marius, C., 1985. Mangroves du Sénégal et de la Gambie: écologie, pédologie, géochimie, mise en valeur et aménagement. Paris, Orstom, coll. Travaux et Documents 193, 357 pp.

Marius, C., 1995. Effets de la sécheresse sur l'évolution des mangroves du Sénégal et de la Gambie, Sécheresse 6: 123–125.

Rodney, W., 1970. A history of the Upper Guinea Coast 1545–1800, Oxford, Clarendon Press 14: 284 pp.

Rollet, B., 1981. Bibliography on mangrove research (1600–1975). Paris, United Nations Educational, Scientific and Cultural Organization (UNESCO): 479 pp.

Ruë, O., 1998. L'aménagement du littoral de Guinée (1945–1995). Mémoires de mangroves, Paris, L'Harmattan: 312 pp.

Sow, M., A., Diallo, N., Diallo, C. A., Dixon, A. & Guissé, 1994. Formations végétales et sols dans les Mangroves des Rivières du Sud. In Cormier-Salem, M. C. (ed.), Dynamique et usages de la mangrove dans les pays des Rivières du Sud (Du Sénégal à la Sierra Leone). Paris, Orstom, coll. Colloques et Séminaires: 353 pp.

Williams, M. (ed.), 1990. Wetlands: A Threatened Landscape. Oxford, Basil Blackwell: 419 pp.

GPSR Compliance

The European Union's (EU) General Product Safety Regulation (GPSR) is a set of rules that requires consumer products to be safe and our obligations to ensure this.

If you have any concerns about our products, you can contact us on ProductSafety@springernature.com

In case Publisher is established outside the EU, the EU authorized representative is:

Springer Nature Customer Service Center GmbH
Europaplatz 3
69115 Heidelberg, Germany

Batch number: 10800215

Printed by Printforce, the Netherlands